Modern Optimisation Techniques in Power Systems

International Series on
MICROPROCESSOR-BASED AND INTELLIGENT SYSTEMS ENGINEERING

VOLUME 20

Editor

Professor S. G. Tzafestas, *National Technical University, Athens, Greece*

Editorial Advisory Board

Professor C. S. Chen, *University of Akron, Ohio, U.S.A.*
Professor T. Fokuda, *Nagoya University, Japan*
Professor F. Harashima, *University of Tokyo, Tokyo, Japan*
Professor G. Schmidt, *Technical University of Munich, Germany*
Professor N. K. Sinha, *McMaster University, Hamilton, Ontario, Canada*
Professor D. Tabak, *George Mason University, Fairfax, Virginia, U.S.A.*
Professor K. Valavanis, *University of Southern Louisiana, Lafayette, U.S.A.*

The titles published in this series are listed at the end of this volume.

Modern Optimisation Techniques in Power Systems

edited by

YONG-HUA SONG
*Department of Electrical Engineering and Electronics,
Brunel University,
West London, U.K.*

KLUWER ACADEMIC PUBLISHERS
DORDRECHT / BOSTON / LONDON

A C.I.P. Catalogue record for this book is available from the Library of Congress.

ISBN 0-7923-5697-7

Published by Kluwer Academic Publishers,
P.O. Box 17, 3300 AA Dordrecht, The Netherlands.

Sold and distributed in North, Central and South America
by Kluwer Academic Publishers,
101 Philip Drive, Norwell, MA 02061, U.S.A.

In all other countries, sold and distributed
by Kluwer Academic Publishers,
P.O. Box 322, 3300 AH Dordrecht, The Netherlands.

Printed on acid-free paper

All Rights Reserved
© 1999 Kluwer Academic Publishers
No part of the material protected by this copyright notice may be reproduced or
utilized in any form or by any means, electronic or mechanical,
including photocopying, recording or by any information storage and
retrieval system, without written permission from the copyright owner

Printed in the Netherlands.

CONTENTS

Preface ix
Contributors xi

Chapter 1 Introduction 1
Y.H. Song

1.1 Power system optimisation 1
1.2 Emerging optimisation techniques 4
1.3 Applications in power systems 12
1.4 References 12

Chapter 2 Simulated annealing applications 15
K. Nara

2.1 Introduction 15
2.2 What is simulated annealing 16
2.3 Power system applications 22
2.4 Concluding remarks 36
2.5 References 36

Chapter 3 Tabu search application in fault section estimation and state identification of unobserved protective relays in power system 39
F. Wen, C.S. Chang

3.1 Introduction 39
3.2 The overall approach for FSE-SIUPR 42
3.3 The mathematical model for FSE-SIUPR 42
3.4 Tabu search with application to FSE-SIUPR 48
3.5 Test results 53
3.6 Conclusions 55
3.7 Appendix 55
3.8 References 59

Chapter 4 Genetic algorithms for scheduling generation and maintenance in power systems 63
C.J. Aldridge, K.P. Dahal, J.R. McDonald

4.1 Genetic algorithm 63
4.2 A knowledge-based genetic algorithm for Unit Commitment 68
4.3 Generator maintenance scheduling using a genetic algorithm 77
4.4 Acknowledgements 86
4.5 References 86

Chapter 5 Transmission network planning using genetic algorithms 91
M.R. Irving, H.M. Chebbo, S.O. Orero

5.1 Introduction 91
5.2 Genetic algorithm 91
5.3 The transmission network planning problem 103
5.4 Experimental results 107
5.5 Future work 109
5.6 Conclusions 109
5.7 Acknowledgement 110
5.8 References 110

Chapter 6 Artificial neural networks for generation scheduling 113
M.P. Walsh, M.J. O'Malley

6.1 Introduction 113
6.2 Hopfield networks 113
6.3 Hopfield networks applied to the scheduling problem 116
6.4 Augmented Hopfield network 123
6.5 Applications of the augmented Hopfield network to the scheduling problem 125
6.6 Other neural network approaches to the scheduling problem 135
6.7 Conclusions 137
6.8 References 137

Chapter 7 Decision making in a deregulated power environment based on fuzzy sets 141
S.M. Shahidehpour, M.I. Alomoush

7.1 Fuzzy sets: introduction and background 141
7.2 Fuzzy multiple objective decision 142
7.3 Analytical hierarchy process 145
7.4 Applications of fuzzy set 148

7.5 Decision making in a deregulated power environment based on fuzzy set … 154
7.6 References … 168

Chapter 8 Lagrangian relaxation applications to electric power operations and planning problems … 173
A.J. Conejo, J.M.Arroyo, N. Jiménez Redondo, F.J. Prieto

8.1 Introduction … 173
8.2 Lagrangian relaxation … 173
8.3 Augmented Lagrangian decomposition … 181
8.4 Short-term hydro-thermal co-ordination and unit commitment … 183
8.5 Decentralized optimal power flow … 189
8.6 Medium /long-term hydro-thermal co-ordination problem … 194
8.7 References … 200

Chapter 9 Inter point methods and applications in power systems … 205
K Xie, Y.H. Song

9.1 Introduction … 205
9.2 Interior point methods … 205
9.3 Power system applications … 220
9.4 Implementation issues … 232
9.5 A case study on IPM based spot pricing alogrithm … 236
9.6 Conclusions … 241
9.7 Acknowledgement … 241
9.8 References … 241

Chapter 10 Ant colony search, advanced engineered-conditioning genetic algorithms and fuzzy logic controlled genetic algorithms: economic dispatch problems … 247
Y.H. Song, C.S.V. Chou, I.K. Yu, G.S. Wang

10.1 Introduction … 247
10.2 Ant colony search algorithms … 247
10.3 Engineered-conditioning GAs … 251
10.4 Fuzzy logic controlled genetic algorithms … 254
10.5 Test results on economic dispatch problems … 255
10.6 Some remarks … 258

10.7 References 259

Chapter 11 Industrial applications of artificial intelligence techniques
A O. Ekwue 261

11.1 Introduction 261
11.2 Algorithmic methods versus artificial intelligence techniques 262
11.3 Artificial intelligence techniques 263
11.4 Conclusions 270
11.5 Acknowledgement 271
11.6 References 272

PREFACE

The electric power industry is currently undergoing an unprecedented reform. The deregulation of electricity supply industry has introduced new opportunity for competition to reduce the cost and cut the price. It is a tremendous challenge for utilities to maintain an economical and reliable supply of electricity in such an environment. Faced by an increasingly complicated existence, power utilities need efficient tools and aids to ensure that electrical energy of the desired quality can be provided at the lowest cost. The overall objective, both for short-term and long-term operations, is then to find the best compromise between the requirements of security and economy. That is, effective tools are urgently required to solve highly constrained optimisation problems.

In recent years, several major modern optimisation techniques have been applied to power systems. A large number of papers and reports have been published. In this respect, it is timely to edit a book on this topic with an aim to report the state of the art development internationally in this area. Chapter 1 introduces the subject of the book by commenting on major modern optimisation techniques covered in the book. They are: simulated annealing, tabu search, genetic algorithms, neural networks, fuzzy programming, Lagrangian relaxation, interior point methods, ant colony search and hybrid techniques. From chapters 2 to 10, detailed descriptions of these modern optimisation techniques together with various power system applications are presented, which clearly demonstrate the potential and procedures of applying such techniques in solving complex power system optimisation problems. The final chapter, Chapter 11, gives an industrial view on applications of some of the techniques. The book is structured so that it is useful to a range of readers, covering basic algorithms as well as applications and case studies.

The Editor would like to thank the authors who submitted their work for inclusion in the book, and Catherine Murphy of Kluwer Academic Publishers for the help in the production of the book.

Professor Yong-Hua Song
Brunel University, London, UK

CONTRIBUTORS

Y.H. Song, M.R. Irving, K. Xie, H. M. Chebbo, S. O. Orero, C.S.V. Chou, I.K. Yu and G.S. Wang
Dept. Of Electrical Engineering
Brunel University
Uxbridge
UB8 3PH, UK

K. Nara
Dept. of System Engineering
Ibaraki University
12-1 Nakanarusawa 4 Chome
Hitachi 316-8511 Japan

Fushuan Wen
Dept. of Electrical Engineering
Zhejiang University
Hangzhou, 310027
Zhejiang Province
P.R.China

C.S. Chang
Dept. of Electrical Engineering
National University of Singapore
10 Kent Ridge Crescent
Singapore 119260
Republic of Singapore

C.J. Aldridge, K.P. Dahal, J.R. McDonald
Centre for Electrical Power Engineering
University of Strathclyde,
Glasgow, G11XW, UK

Michael P. Walsh And Mark J. O'Malley
Department of Electrical Engineering
University College, Dublin,
Ireland

S. M. Shahidehpour And M. I. Alomoush
Department of Electrical and Computer Engineering
Illinois Institute of Technology
Chicago, IL 60616

A.J. Conejo And J.M.Arroyo
ETSII, Univ. Castilla–La Mancha
Ciudad Real, 13071, Spain

N. Jiménez Redondo
Univ. De Málaga
Málaga, Spain

F.J. Prieto
Univ. Carlos Iii
Madrid, Spain

A.O. Ekwue
National Grid Company plc,
RG41 5BN, UK

Chapter 1

INTRODUCTION

Y.H. Song
Department of Electrical Engineering and Electronics
Brunel University, Uxbridge
UB8 3PH, UK

1.1 Power System Optimisation

In electricity supply systems, there exist a wide range of problems involving optimisation process [1, 2]. It may include individual generation, transmission, distribution systems or any combination of them. In general, the target is to minimize the costs for construction and operation of the system. The economic efficiency or profits thus becomes the objective function and the other requirements are represented by the constraints. The decisions concerned can cover periods of different lengths: long-term, medium term, short term and on-line. This forms a hierarchical structure typically from expansion planning, maintenance scheduling, fuel resource scheduling, unit commitment, load dispatch to optimal power flow. Recent introduction of deregulation into electricity supply industry adds new dimension in such optimisation problems with the maximum market benefit as its objective [3].

1.1.1 A GENERAL FRAMEWORK

Mathematically, many real world power system optimisation problems can be formulated generally as follows:

Minimise $f(x)$
Subject to $g(x) = 0$
$\qquad h(x) \leq 0$

where x is the decision variable; $f(x)$ is a real-valued objective function; $g(x)$ and $h(x)$ represent the equality constraints and inequality constraints respectively. The points x satisfying the constraints are called the feasible solutions. A point x^* is globally optimal solution if $f(x^*) \leq f(x)$ for all x in the feasible region. It is a locally optimal solution if $f(x^*) \leq f(x)$ for the x in a neighbourhood of x^*.

There are different classes of such problems, depending on the way the function and constraints are specified. For example, the decision variables involved can be discrete or continuous. The functions can be linear or nonlinear. The most well-known class is that obtained by restricting the functions to be linear function and decision variables continuous. This is so called linear programming. On the other hand, combinatorial optimisation refers to problems in which the decision variables are discrete.

As examples, the following illustrate several typical power system optimisation problems.

1.1.2 CLASSIC ECONOMIC DISPATCHING

The basic requirement of power economic dispatch (ED) is to generate adequate electricity to meet continuously changing customer load demand at the lowest possible cost under a number of constraints.

The objective of the ED problem is to minimize the total fuel cost at thermal plants:

Subject to the following constraints:

(a) The real power balance
(b) The real power limits on the generator outputs

1.1.3 UNIT COMMITMENT

Unit commitment is the problem of determining the schedule of generating units within a power system subject to equipment and operating constraints. The decision process is to select units to be on or off. The generic unit commitment problem can be formulated as:

The objective function to be minimised the fuel costs, the maintenance costs and the start up costs.

Subject to the following constraints:

(a) real power balance constraint
(b) real power operating limits of generating units
(c) ramp rate limits of generating units
(d) line flow limits of transmission line
(e) spinning reserve constraint
(f) minimum up time of units
(g) minimum down time of units

The unit commitment problem is a mixed integer nonlinear programming problem.

1.1.4 OPTIMAL POWER FLOW

The principal goal of the Optimal Power Flow is to provide the electric utility with suggestions to optimize the current power system state online with respect to various objectives under various constraints. Depending on the specific objectives and constraints, there are different OPF formulations. Most current interest in the OPF centres around its ability to solve for the optimal solution that takes account of the security of the system.

The typical objectives are minimization of the total generation cost, minimization of the active power losses, maximization of the degree of security of a network, minimization of shift of generation and other controls, minimization of load shedding schedule under emergency conditions or a combination of some of them. The ability to use different objective functions provides a very flexible analytical tool.

A large number of equality and inequality constraints can be included in the OPF formulation. In addition to the power flow constraints and active power generation limits in the aforementioned economic dispatch problem, there are additional operational constraints which can be included. These are limits on the generator reactive power, limits on the voltage magnitude at generation and load buses, flows on transmission lines or transformers, and limits on control variables. The OPF can also include constraints that represent operation of the system after contingency outages.

Furthermore, there are many more control variables in the OPF, which provide a powerful means to optimise the operation of the power system.

In view of these, the OPF is a very large and very difficult mathematical programming problem.

1.1.5 OPERATION OF POWER SYSTEMS IN A COMPETITIVE ENVIRONMENT

The utility industry restructuring has enhanced the role and importance of OPF tools. Some of the basic business functions can not be performed without OPF. A robust and flexible Optimal Power Flow program will be the core of so called "Price Responding Energy Management System" in the emerging market structure. Moreover, market requirements for pricing nodal injection and other services and constraints can introduce new elements and complexity to the OPF formulation. In general, the objective of an electricity market is to maximize market benefits. This is equivalent to minimizing the sum of payments to energy/reserve offers and the revenues from demand bids.

1.2 Emerging Optimisation Techniques

A large number of optimisation techniques have been applied to power systems. These include: Dynamic programming; Integer and mix-integer programming; Branch-and-bound; Linear programming; Network flow programming; and Newton methods etc. Over the past three decades, there has been a burgeoning interest in heuristic search methods for complex optimisation problems [4, 5]. It is the intention of the book to cover some power system applications of these emerging optimisation techniques. In this section, an overview is presented.

1.2.1 A GENERAL SEARCH PARADIGM

Various attempts have been made to produce unified approaches to optimisation through the use of guided search techniques. The following illustrates the procedure of a general search paradigm.

```
P,Q,R: (multi-)set of solutions S
initialise (P);
while not finish(P) do
begin

    Q:= select(P)
    R:= create(Q)
    P:= merge(P,Q,R)

end
end
```

P is the pool of potential solutions(>=1). In some cases, such as simple Genetic Algorithms, S may contain multiple copies of a solution (a multiset), or in other cases, S only contains single copies (set). Q is the subpool selected and used to create new sets of solutions. R is the set of new solutions. In different search algorithms, the "select," "create" and "merge" will be in different forms.

The root of many modern optimisation techniques hinges on the idea of neighbourhood search (also called local search). It essentially involves search through the solution space by moving to the neighbours of a known solution in a direction such as the steepest ascent. The process is repeated at the new point and the algorithm continues until a local minimum is found. Obviously, the major problem is that the solution it generates is usually only a local optimum.

1.2.2 SIMULATED ANNEALING

Simulated annealing techniques were originally inspired by the formation of crystals in solids during cooling. As Iron Age blacksmiths discovered, the slower the cooling, the more perfect the crystals that form. The reason for this is that the randomness provided by thermal energy allows atoms to escape from locally-optimal configurations (small sub-crystals) and form globally-optimal configurations instead. In the mid-1970s, Kirkpatrick et al used an analogy to annealing to optimize the layout of printed circuit boards. In essence, SA reduces the chance of getting stuck in a local optimum by allowing moves to inferior solutions under the control of a randomised strategy.

In SA [6], Each iteration consists of the random selection of a configuration from the neighbourhood of the current configuration and the calculation of the corresponding change in cost function . The neighbourhood is defined by the choice of a generation mechanism, i.e. a "prescription" to generate a transition from one configuration into another by a small perturbation. If the change in cost function is negative, the transition is unconditionally accepted; if the cost function increases the transition is accepted with a probability based upon the Boltzmann distribution. This temperature is gradually lowered by the so called cooling rate throughout the algorithm from a sufficiently high starting value (i.e. a temperature where almost every proposed transition, both positive and negative, is accepted) to a "freezing" temperature, where no further changes occur. In practise, the temperature is decreased in stages, and at each stage the temperature is kept constant until thermal quasi-equilibrium is reached. The whole of parameters determining the temperature decrement (initial temperature, stop criterion, temperature decrement between successive stages, number of transitions for each temperature value) is called the cooling schedule.

The advantages of Simulated annealing are its general applicability to deal with arbitrary systems and cost functions, its ability to statistically guarantees finding an optimal solution, and its simplicity of implementation, even for complex problems. This makes SA an attractive option for optimization problems where specialized or problem specific methods are not available. The major drawback is that repeatedly annealing with a schedule is very slow. For problems where the energy landscape is smooth, simpler and faster local methods will work better. The method cannot tell whether it has found an optimal solution. Some other method (e.g. branch and bound) is required to do this. In this respect, SA is normally used as an approximation algorithm.

1.2.3 TABU SEARCH

The roots of TS [7] go back to the Glover's work in 1970s. Tabu Search (TS) is an iterative improvement procedure that starts from some initial feasible solution and attempts to determine a better solution in the manner of a 'greatest descent neighbourhood' search algorithm. It escapes local optima by imposing restrictions, using a short-term memory of recent solutions and strategies implied from long-term memory processes, to guide the search process.

At TS, the neighbourhood, which is being used to generate a subset of neighbours from which to select the next solution/move, is modified by classifying some moves as tabu, others as desirable. This is the key element of TS and is called tabu list management. In other words, Tabu list management concerns updating the tabu list, i.e., deciding on how many and which moves have to be set tabu within any iteration of the search. There are in several basic ways for carrying out this management, generally involving a recency based record that can be maintained individually for different attributes or different classes of attributes. In addition, many applications of tabu search introduce memory structures based on frequency (modulated by a notion of move influence), and the coordination of these memory elements is made to vary as the preceding short term memory component becomes integrated with longer term components. The purpose of this integration is to provide a balance between two types of globally interacting strategies, called intensification strategies and diversification strategies.

A distinguishing feature of TS, represented by its exploitation of adaptive forms of memory, equips it to penetrate complexities that often confound alternative approaches. Tabu search is an engineering approach that must be tailored to the details of the problem at hand. Unfortunately, there is little theoretical knowledge that guides this tailoring process, and users have to resort to the available practical experience.

1.2.4 GENETIC ALGORITHM

Genetic algorithms were invented by Holland to mimic some of the processes of natural evolution and selection. In nature, each species needs to adapt to a complicated and changing environment in order to maximise the likelihood of its survival. The knowledge that each species gains is encoded in its chromosomes which undergo transformations when reproduction occurs. Over a period of time, these changes to the chromosomes give rise to species that are more likely to survive, and so have a greater chance of passing their improved characteristics on to future generations. Of course, not all changes will be beneficial but those which are not tend to die out. Several different types of evoluationary search methods were developed independently. These include (a) genetic programming (GP), which evolve programs, (b) evolutionary programming (EP), which focuses on optimizing continuous functions without recombination, (c) evolutionary strategies (ES), which focuses on optimizing continuous functions with recombination, and (d) genetic algorithms (GAs), which focuses on optimizing general combinatorial problems.

A simple genetic algorithm [8] involves the following steps: (1) Code the problem; (2) Randomly generate initial population strings; (3) Evaluate each string's fitness; (4) Select highly-fit strings as parents and produce offsprings according to their fitness; (5) Create new strings by mating current offsprings, apply crossover and mutation operators to introduce variations and form new strings; (6) Finally, the new strings replace the existing ones. This sequence continues until some termination condition is reached. In this process, the crossover and mutation operators play a key role.

GAs differ from traditional optimization algorithms in several important respects: (1) They work using an encoding of the control variables, rather than the variables themselves. (2) They search from one population of solutions to another, rather than from individual to individual. (3) They use only objective function information, not derivatives. It must be noted that standard GAs are often found to experience convergence difficulties. In many applications GAs locate the neighborhood of the global optimum extremely efficiently but have problems converging onto the optimum itself. Other difficulties with the standard GAs are the computation efficiency and premature convergence.

1.2.5 NEURAL NETWORKS

The publication of Hopfield's seminal paper in 1982 started the modern era in neural networks [9]. The Hopfield network is a recurrent network that embodies a profound physical principle, namely, that of storing information in a dynamically stable configuration. Hopfield's idea is to locate each pattern to be stored at the bottom of a valley of an energy landscape, and then permitting a dynamical procedure to minimise the energy of the network in such a way that the valley becomes a basin of attraction.

Essentially, in applying Hopfield networks for solving power system optimisation problems, the energy function of the Hopfield network has to be made equivalent to a certain objective function that needs to be minimised. The search for a minimum energy performed by the Hopfield network corresponds to the search for a solution to the optimisation problem. The constraints are mapped into the network by taking feedbacks from variable neurones (computing the optimisation variables) to the constraint neurones (computing the constraint functions). Thus, the key is to formulate the power system optimisation problem with its objective function which can be used to construct a Hopfield network, i.e. the weights. Once the problem is formulated in this way, a particular type of parallel algorithm has been constructed. It is clear that the formulation is application-oriented.

1.2.6 ANT COLONY SEARCH

Ant Colony Search (ACS) studies artificial systems that take inspiration from the behavior of real ant colonies and which are used to solve function or combinatorial optimization problems. Currently, most work has been done in the direction of applying ACS to combinatorial optimization. The first ACS system was introduced by Marco Dorigo in his Ph.D. thesis (1992), and was called Ant System. Ant colony search algorithms, to some extent, mimic the behavior of real ants. As is well known, real ants are capable of finding the shortest path from food sources to the nest without using visual cues. They are also capable of adapting to changes in the environment, for example, finding a new shortest path once the old one is no longer feasible due to a new obstacle. The studies by ethnologists reveal that such capabilities ants have are essentially due to what is called "pheromone trails" which ants use to communicate information among individuals regarding path and to decide where to go. Ants deposit a certain amount of pheromone while walking, and each ant probabilistically prefers to follow a direction rich in pheromone rather than a poorer one.

In ACS [10], the colony consists of many homogeneous ants communicating among them by recruit pheromone. Ants change their behaviours according to the situation. Firstly, they walk randomly to search the bait sites in an environment of discrete time. They will not be completely blind, decision is made based on the intensity of trace perceived and the visibility. When the bait site is found, they will carry bait from the site to the nest secreting the pheromone. Also, each of them will have some memory about their location and the next possible move. In ACS, according to the objective function, their performance will be weighed as fitness value, the direct influence of which is on the level of trail quantity adding to the particular directions the ants have selected. Each ant chooses the next node to move taking into account two parameters: the visibility of the node and the trail intensity of trail previously laid by other ants.

The main characteristics of ACS are positive feedback, distributed computation and the use of a constructive greedy heuristic. Positive feedback accounts for rapid discovery of good solutions, distributed computation avoids premature convergence, and the greedy heuristic helps to find acceptable solutions in the early stages of the search process. More potentially beneficial work remains to be done, particularly in the areas of improvement of its computation efficiency.

1.2.7 FUZZY PROGRAMMING

Fuzzy logic was developed by Lotfi Zadeh in 1964 to address uncertainty and imprecision which widely exist in engineering problems. The complexity in the world generally arises from uncertainty in the form of ambiguity. In real-world problems, there exist uncertainties in both the objective functions and constraints. By using fuzzy set theory [11], these uncertainties can be considered. For example, multiple objectives can be reformulated as fuzzy sets. The membership value indicates the degree to which an objective is satisfied; the higher the membership, the greater the satisfaction with a solution. Constraints are also modelled by fuzzy sets. The benefits are: (1) the fuzzy sets more accurately represent the operational constraints of the power system; and (2) Fuzzified constraints will be softer than traditional ones, i.e. the resulting solutions will tend to be well within the crisp constraint regions.

The salient feature of fuzzy set theory is the ability to tackle "fuzzy" information. The best application in the areas of optimisation is to use it as an interface between a real-world problem and other optimisation techniques.

1.2.8 LANGRANGIAN RELAXATION

Lagrangian relaxation was developed in the early 1970s with the pioneering work of Held and Karp on the travelling salesman problem. The basic idea of Lagrangian relaxation is to relax the constraints by using Lagrangian multipliers. The relaxed problem is then decomposed into a number of subproblems. The search process is an iterative algorithm that solves relaxed subproblems and updates Lagrangain multipliers according to the extent of violation of constraints.

There are two basic reasons for the successful applications of LR: (1) Many real-world problems are complicated by the addition of some constraints. The LR absorbs these complicated constraints into the objective function. (2) Practical experience with LR has indicated that it performs well at reasonable computation cost.

1.2.9 INTERIOR POINT METHODS

With the 1984 publication of his paper announcing a polynomial-time algorithm for linear programming, Karmarkar started an amazing flurry of activity in the area of interior-point linear programming. Karmarkar's original method involved the use of a projective scaling which served to move the current solution point to the center of the polytope in order to allow "long" steps toward the optimum. Interior-point algorithms get their name because the search directions generated strike out into the interior of the polytope rather than skirting around the boundaries as do their simplex-like cousins. Interior-point research has gone beyond the scope of just linear programming. In 1986, Gill et al showed the relationship between Karmarkar's algorithm and the so-called "logarithmic barrier function algorithm."

Most power system applications employ a primal-dual logarithmic barrier technique, which eliminates all inequality constraints from the problem by adding a term to the objective function. These terms are called "barrier functions," because they form an infinite barrier, or "wall," around the boundaries of the feasible region. Unlike penalty functions used by other methods, the barrier function treats all limits as hard constraints, permitting no violation whatsoever unless directed to do so by the user in the event of infeasibility. By replacing the inequality constraints with barrier functions, the algorithm is able to converge onto the correct set of binding inequality constraints while at the same time converging to a zero mismatch solution.

The interior-point methods [12] tend to converge in many fewer iterations than other methods, though each iteration is much more expensive.

1.2.10 HYBRID TECHNIQUES

Due to the complexity of real-world problems and the pro and cons of various search techniques, it is apparent that hybridisation is a way forward to develop more powerful algorithms. Hybridisation allows searches that display particular properties to be produced. The development of hybrids and their theoretical underpinning is a new area of research that is just starting to be explored. Some of the emerged hybrid techniques [14-24] are briefly presented here to appreciate the advantages of such techniques.

(i) Integrating SA and TS

A number of differences can be identified between SA and TS. First, TS emphasises scouting successive neighbourhoods to identify high quality moves, while SA randomly samples from neighbourhoods. Second, TS evaluates the relative attractiveness of moves in relation to both the fitness function and influence factors. Finally, TS relies on guiding the search by use of multiple thresholds, SA uses only one (temperature). Hybrids that allow temperature to be strategically manipulated in SA have been shown to produce improved performance over standard SA approaches. In SA moves are usually blind. One approach to integrating intelligent moves is to add probabilistic TS with attraction and spatial memory. A hybrid method that expands the SA basis for move evaluations also has been found to perform better.

(ii) Integrating SA and GA

A number of authors have attempted to integrate the SA and GA approaches. The SA community has borrowed elements from GAs in an attempt to parallelise SAs. The GA community uses SA concepts as a randomising element. Parallel Recombinative Simulated Annealing is a parallelised version of SA that attempts to bring the gains of the implicit parallelism of GAs to the search of the solution space. It is a hybrid with a population of n solutions and may prove useful in combinatorial optimisation problems, potentially allowing the allocation problem to be extended to include changes in infrastructure, such as extra processors.

Another integration is the COoperative Simulated Annealing which inherits the idea of population and information exchange from Genetic Algorithms but replaces the usual crossover by the so-called cooperative transitions. The acceptance probability for the transition is controlled by the Metropolis function, depending on the virtual variable called „temperature" and the difference in the solutions' goal function values (as usually done in SA).

(iii) Integrating GA and local search

Methods to overcome problems associated with GA are often conflicting and a compromise is usually required. Attempts have been made to use problem-specific knowledge to direct the search. Often the hybrids produced incorporate elements of other neighbourhood search techniques. A number of theoretical objections have been raised about hybrid GAs. However, in practice, hybrid GAs do well at optimisation tasks.

Engineering-conditioning GA is a hybrid between GAs and Hill Climbing. The search is undertaken in two stages: a clustering stage and a tuning stage when the population is near convergence. The primary hybridisation is to introduce a simple problem-specific hill-climbing procedure that can increase the fitness of individuals.

Another interesting integration is the hybridisation of GA with Lagrangian relaxation methods.

(iv) Integrating fuzzy logic and GA

In recent years, various techniques have been studied to improve genetic search. Particularly, as genetic algorithms are distinguished from others by the emphasis on crossover and mutation, more recently much attention and effort has been devoted to improving them. In this respect, two-point, multi-point and uniform crossover, and variable mutation rate have been recently proposed. More advanced genetic operators have been proposed which are based on fuzzy logic with the ability to adaptively/dynamically adjust the crossover and mutation during the evolution process. For example, the heuristic updating principles of the crossover probability is if the change in average fitness of the populations is greater than zero and keeps the same sign in consecutive generations, then the crossover probability should be increased. Otherwise the crossover probability should be decreased.

1.3 Applications in Power Systems

The aforementioned optimisation techniques have being applied to a variety of power system optimisation problems with the major effort on scheduling and operation optimisation. Several good review papers [25-20] have appeared recently either surveying power system applications of a particular group of techniques or various optimisation techniques applied to a particular power system problem. It is very difficult, if not impossible, to give an exhaustive treatment of such a growing area in power systems, but an attempt has been made in this book to cover most of the major developments in recent years.

1.4 References

1. A J Wood, B F Wollenberg, Power Generation Operation and Control, *John Wiley & Sons*, New York, 1996
2. K Frauendorfer, H Glavitsch, R Bacher, Optimisation in Planning and Operation of Electric Power Systems, *Physica-Verlag*, Germany, 1993
3. M Ilic, F Galiana, L Fink, Power Systems Restructuring, *Kulwer Academic Publishers,* Boston, 1998
4. C R Reeves, Modern Heuristic Techniques for Combinatorial Problems, *McGraw-Hill*, London, 1995
5. E Aarts, J K Lenstra, Local Search in Combinatorial Optimisation, *John Wiley & Sons*, 1997
6. E Aarts, J Korst, Simulated Annealing and Boltzmann Machines, *John Wiley & Sons*, 1989
7. F Glover, M Laguna, Tabu Search, *Kluwer Academic Publishers*, 1997
8. D A Goldberg, Genetic Algorithms in Search, Optimisation and Machine Learning, *Addison-Wesley*, 1989
9. S Haykin, Neural Networks, *Macmillan College Publishing Company*, 1994
10. M Dorigo, V Maniezzo and A Colorni, The Ant System: Optimization by a Colony of Co-operating Agents, *IEEE Transactions on Systems, Man and Cybernetics*, 1995
11. T J Ross, Fuzzy Logic with Engineering Applications, *McGraw-Hill*, 1995
12. S J Wright, Primal-Dual Interior-Point Methods, *SIAM*, 1996
13. Y H Song, A T Johns, Applications of fuzzy logic in power systems, Part 1: General Introduction to Fuzzy Logic, *IEE Power Engineering Journal*, Vol.11, No.5, 1997, pp.219-222
14. R K Aggarwal, Y H Song, Artificial neural networks in power systems: Part 1 General introduction to neural computing, *IEE Power Engineering Journal*, Vol.11, No.3, 1997, pp.129-134
15. B Fox, Integrating and Accelerating Tabu Search, Simulated Annealing and Genetic Algorithms, *Annals of OR*, 41:47-67, 1993

16. D. Goldberg and S. Mahfoud. Parallel Recombinative Simulated Annealing: A Genetic Algorithm, *Technical Report 92002*, Illinois Genetic Algorithms Laboratory, 1993
17. Wendt, Oliver: COSA: COoperative Simulated Annealing - Integration von Genetischen Algorithmen und Simulated Annealing am Beispiel der Tourenplanung; Forschungsbericht *94-09 des Instituts für Wirtschaftsinformatik*, 1994
18. F Lin, C Kao, and C Hsu. Applying the Genetic Algorithm to Simulated Annealing in Solving NP-Hard Problems, *IEEE Trans. on Systems, Man &Cybernetics*, 23(6):1752-1767, 1993
19. Y H Song, C S Chou, Advanced engineered-conditioning genetic approach to power economic dispatch, *Proc IEE - GTD*, Pt.C, Vol.144, No.3, 1997, pp.285-292
20. S O Orero, M R Irving, A combination of the genetic algorithm and Lagrangian relaxation decomposition techniques for the generation unit commitment problem, *Int. Jnl. Of Electric Power Systems Research*, 1997
21. Y H Song, R Dunn, Fuzzy logic and hybrid systems, in Artificial Intelligence Techniques in Power Systems, Edited by K Warwick, A Ekwue, R Aggarwal, 1997, *The IEE*
22. G S Wang, Y H Song, A T Johns, P Y Wang, Z Y Hu, Fuzzy logic controlled learning algorithm for training multilayer feedforward neural networks, *International Journal of Neural Network World*, No.1, 1997, pp.73-88
23. Y H Song, G S Wang, A T Johns, P Y Wang, Distribution network reconfiguration for loss reduction using fuzzy controlled evolutionary programming, *Proc IEE - GTD*, Pt.C, Vol.144, No.4, 1997, pp.345-350
24. Y H Song, G S Wang, P Y Wang, A T Johns, Enviromental/economic dispatch using fuzzy controlled genetic algorithms, *Proc IEE - GTD*, Pt.C, Vol.144, No.4, 1997, pp.377-384
25. R J Sarfi, M M Salama, A Y Chikhani, A Survey of the Sate of the Art in Distribution System Reconfiguration for System Loss Reduction, *Electric Power System Research*, 1994
26. G B Sheble, G N Fahd, Unit Commitment Literature Synopsis, *IEEE Trans Power Systems*, Vol.9, No.1, 1993, pp.128-135
27. B H Chowdhury, S Rahman, A Review of Recent Advances in Economic Dispatch, *IEEE Trans Power Systems*, Vol.5, No.4, 1990, 1248-12555
28. D Srinivasan, F Wen, C S Chang, A C Liew, A Survey of applications of Evolutionary Computing to Power systems, *Proc International Conference on Intelligent System Application to Power Systems*, Orlando, USA, 1996
29. Y H Song (Guest Editor), Special Issue on Fuzzy Logic Applications in Power System, *International Journal of Fuzzy Sets and Systems*, 1998
30. Y H Song, M El-Sharkawi, M Mori, (Guest Editors), Special Issue on Neural Networks in Power Systems, *International Journal of Neural Computing*, 1998

Chapter 2

SIMULATED ANNEALING APPLICATIONS

K. Nara
Ibaraki University
12-1 Nakanarusawa 4 Chome
Hitachi 316-8511 JAPAN

2.1 Introduction

Simulated Annealing (SA) is a method to solve an optimization problem by simulating a stochastic thermal dynamics of a metal cooling process. SA obtains an optimal solution by simulating a physical fact that liquid metal transmutes to be crystal (which has the smallest internal thermal energy) if it is cooled satisfactory slowly from a high temperature state (with large internal thermal energy). In other words, functional minimization corresponds to minimization of internal thermal energy of metal in a melting pot. The idea to solve the optimization problem by using this fact is proposed by Kirkpatric [17]. He demonstrated that the method is successfully applicable to so-called combinatorial optimization problem. In the reference [17], he has introduced several applications (physical design of computers, wiring and traveling salesmen) of SA. In power system applications, many combinatorial optimization problems have been solved by using SA. There are many kinds of combinatorial optimization problems in power systems area: generator maintenance scheduling, unit commitment, VAR planning, network planning, distribution systems reconfiguration etc. It is well known that we have not yet found an established solution algorithm for these problems. Most efficient solution algorithms proposed for the problems can only find approximated optimal solutions, and we cannot find the exact optimal solution for the practical problem. It is well known that SA can find near optimal solutions for these combinatorial optimization problems if we decrease temperature carefully and slowly enough. Moreover, the simulated annealing algorithm is so simple as shown in later sections. Therefore, solutions obtained by using SA are often used as a reference solution to evaluate the performance of newly developed solution algorithms even though it takes huge computational burden.

In this Chapter, in section 2.2, basic theory of the simulated annealing is explained. In section 2.3, applications of the method to power systems planning and operation are shown. Parallel computation method of SA and combined use of SA with other mathematical methods (or modern heuristics) are also explained in the following sections.

2.2 What is Simulated Annealing?

2.2.1 ANNEALING PROCESS

When metal is heated up in a melting pot, its internal thermal energy increases, and its phase turn to be a liquid phase. At this state, molecules of metal are in random position and are moving freely in a melting pot with very high speed as shown in Fig.1 (a). In other words, the state of molecules (positions and moving speeds) makes the internal thermal energy high. Since the state or internal energy of a metal is determined stochastically, according to the molecule's behavior, internal energy is large when its temperature is satisfactory high. From this high temperature state, if temperature of a metal is cooled down slowly, its internal thermal energy decreases slowly although it sometimes increases ruled by the Boltzmann's probability,

$$\exp(-\Delta E / k_B T) \tag{1}$$

where, ΔE : internal energy difference between two different states,
  T : absolute temperature
  k_B : Boltzmann constant

When it is cooled down, molecule's moving speed becomes slower according to its internal energy decrease as shown in Fig.1(b). Soon, the phase of metal turn to be solid, and molecule's moving speed becomes slower and slower. Since a cooling process of metal is ruled by stochastic thermal dynamics, the final state (positions) of molecules is determined randomly according to the behavior of the molecules or its cooling speed. If a metal is cooled so fast, final state of a metal is amorphous (glass-like). If a metal is cooled very slowly, the final state of a metal is a ground state and makes crystal (all the molecules line-up) at 0K($-273\,^oC$) as shown in Fig.1(c). A ground state of a metal has the smallest internal thermal energy among any other states it can take.

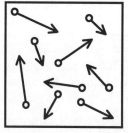 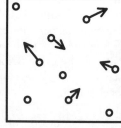 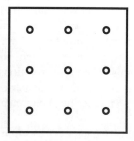

(a) High temperature   (b) Low temperature   (c) Frozen
Fig. 1 Annealing process of metal

The process to heat up a metal and to cool it down slowly is referred to as "annealing". On the other hand, the process to heat up a metal and to cool it down fast is referred to as "quenching" or "hardening".

By simulating a nature of an annealing process in which the smallest internal energy can be reached, we can realize a minimization of an objective function of a complex optimization problem.

2.2.2 DIGITAL SIMULATION OF ANNEALING PROCESS

Annealing process can be simulated by a digital computer if we consider that the process is a behavior of a set of many discrete particles. By using this concept, Metoropolis has succeeded to develop an algorithm to simulate the stochastic process which can realize the thermal equilibrium at some constant temperature T. To develop the algorithm, he substitutes a set of molecule's movement by a series of a small deviation of each molecule. In other words, the process is simulated by a Markov chain of states determined by deviations of discrete particles. The resulting Metoropolis's algorithm is as follows.

Metoropolis's Algorithm

(Step1) Select a new state X_2^t if a current state is X_1^t at time t.

Where, X_2^t is a state randomly created by applying a small deviation to (neighborhood of) X_1^t. Here, the following symmetry condition must be satisfied.

$$p(X_2^t = x_j | X_1^t = x_i) = p(X_2^t = x_i | X_1^t = x_j) \qquad (2)$$

Where, $p(A|B)$: conditional probability, x_i: state of metal.

(Step2) Calculate $\Delta E = E(X_2) - E(X_1)$

Where, $E(X)$ means an internal thermal energy of state X.

(Step3) State transition is determined by the following probability

$$\min \{ 1, \exp(-\Delta E / T) \} \qquad (3)$$

Where, T is a constant which is referred to as "temperature" originated by physical analogy (*cf.* eq.(1)).

In this method, the state always changes to X_2^t if $\Delta E < 0$. If $\Delta E > 0$, then the state move to X_2^t with probability $\exp(-\Delta E / T)$. Therefore, the transition probability of the state ($\Delta E > 0$) is calculated as follows.

$$X_1^{t+1} = X_2^t : \text{Prob.} = \exp(-\Delta E / T) \qquad (4)$$

$$X_1^{t+1} = X_1^t : \text{ Prob.}=1-\exp(-\Delta E/T) \tag{5}$$

The above procedure generates a Markov chain of transition probabilities which converge to a unique Gibbs distribution.

2.2.3 SIMULATED ANNEALING

Metoropolis's algorithm gives us a way to simulate to reach a thermal equilibrium state at one temperature. Therefore, if we try to apply Metoropolis's algorithm as many times as to reach a equilibrium from some high temperature, and continuously decrease a temperature very slowly after the thermal equilibrium has reached in each temperature, then we may obtain an approximate smallest internal energy state when temperature becomes satisfactory low. Here, if the internal energy corresponds to an objective function of optimization problem, we can find the approximate minimal solution of the problem. Kirkpatrik first applied this algorithm to functional optimization [17].

Simulated Annealing Algorithm
Kirkpatrik showed that Metoropolis's algorithm can be applied to the functional optimization by corresponding parameters of the annealing process to parameters of the optimization problem as shown in Table 1. The resulting algorithm is referred to as *Simulated Annealing algorithm*.

Table 1 Relationship between thermal dynamics and functional optimization

Thermal dynamics	Functional optimization
State	Solution
Energy	Cost or Objective function
State transition	Neighborhood solution
Temperature	Control parameter (temperature)
Freezing point	Heuristic solution

The resulting simulated annealing algorithm is shown in Fig.2. In Fig.2, firstly, initial solution and initial temperature are selected. Initial temperature must be satisfactory high although it normally is found by trial and error. Secondly, a trial solution is generated in the neighborhood of the current solution. Then, a difference of the objective functions between current(t) and trial(t+1) solutions

$$\Delta = f(X^{t+1}) - f(X^t) \tag{6}$$

is calculated. If $\Delta <0$, then trial solution is accepted. If $\Delta >0$, then trial solution is accepted with probability $\exp(-\Delta/T^k)$. In a real computer implementation, the

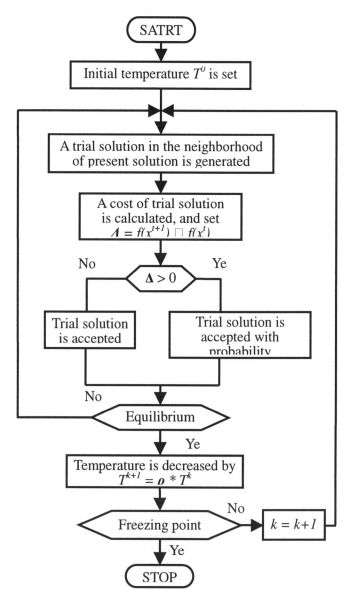

Fig. 2 Algorithm of simulated annealing

solution is accepted when $\exp(-\Delta/T^k) > r$ holds. Where, r is a random number in the range 0 to 1. The above procedures are repeated until the thermal equilibrium of the current temperature reaches. In a real implementation, thermal equilibrium is estimated when a predetermined large number (several hundreds or thousands) of trial solutions are evaluated. When equilibrium reaches, the temperature is decreased by the following equation for instance, and the above processes are repeated with the new temperature until freezing point reaches.

$$T^{k+1} = \rho T^k \tag{7}$$

Where, ρ is taken as 0.80 to 0.99. If the freezing point reaches, the algorithm stops. If we use eq.(7), we cannot reach 0K or the freezing point. Therefore, in a real implementation, we normally determine that we have reached the freezing point if acceptance rate of solutions (when $\Delta > 0$) becomes less than the predetermined value, though temperature does not reach to the absolute zero.

In the real implementation of the above algorithm to solve the optimization problem, we must determine two issues: the generation mechanism of neighborhood solution and the cooling schedule. Since the generation mechanism of a neighborhood solution completely depends on the problem, it will be explained when some real applications are introduced in section 2.3. Here, some discussions of the cooling schedule are shown.

Cooling Schedule
The cooling schedule severely affects the solution efficiency and the quality of the solution. Therefore, in applying a simulated annealing algorithm to an optimization problem, it is important to decrease a temperature satisfactory slowly to obtain a qualified solution. Geman,S. et al. have proved that we can obtain the exact optimal solution if we give trial solutions infinitive times in one temperature, and decrease temperature so slowly as to satisfy the following condition [12].

$$T > c / \log k \tag{8}$$

Where, k means an iteration number of temperature decrease, and c is a constant independent to temperature T. However, the above condition takes an infinitive computation burden. Therefore, in a real computational implementation, more practical cooling schedule is employed. Normally, finite times trial solution is generated in one temperature, and the cooling schedule calculated by eq.(7) is commonly employed. Eq.(7) shows an exponential decrease of temperature, and a large computation time is necessary to attain a satisfactory low temperature. To reduce this computational burden, several cooling schedules are proposed. One of then is

$$T^{k+1} = T^k / (1 + \beta T^k) \tag{9}$$

Where, β is a very small constant. In this cooling schedule, temperature decrease is so slow, but a trial solution in one temperature is limited to one time. Many other improvements of the cooling schedule have been proposed. More detailed discussions of a cooling schedule can be found in other publications (for example,[21])

Comparison with Descent Algorithm
Simulated annealing can be considered as a revised method of a descent algorithm. From an initial solution, a descent method searches the optimal solution only to a descending (down valley) direction in a solution space. By the descent method, a solution is often trapped at a local minimum. SA basically searches the optimal solution also to descending direction. However, it sometimes permits to go to up-hill direction with a probability,

$$\exp(-\Delta E / T) \tag{10}$$

Therefore, solution sometimes jumps a valley as shown in Fig.3, and cannot be trapped in a local minimum. When temperature is high, such a jump occurs so often. If temperature becomes low, solution scarcely moves to the up-hill direction as easily be understood by eq.(10). By jumping to a worse solution, solution can escape from a local minimum, and can finally converge to the global minimum. If we decrease a temperature so fast, the final solution normally be trapped in a shallow valley (local minimum). However, if temperature is decreased slowly enough, the final solution can find the deepest valley (global minimum) through the simulation of a nature of physical phenomenon.

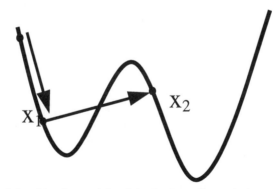

Fig. 3 Searching characteristics of simulated annealing method

Parallel Computation
Although SA can find near optimal solutions for a combinatorial optimization problems, the only deficit is that it takes huge computation time to simulate a

thermal equilibrium in many temperatures trough realizing the slow cooling. To overcome this deficit, parallel computation methods of SA have been proposed. However, since SA is fundamentally based on Markov chain, it is difficult to realize a parallel algorithm. The principle ideas for parallel calculation are as follows.

(1) Each processor evaluates one perturbation of the Markov chain under the condition that the same perturbation cannot be allowed to more than two processors.
(2) A search space is divided into several subspaces, and each subspace is searched simultaneously. Move between each subspace is governed by another processor, and information about moves is broadcast to all processors.
(3) Moves which do not interact each other are assigned to different processors, and are calculated simultaneously.
(4) Different processors calculate different short Markov chains. To obtain a good solution, the concept of Markov chain must be maintained even in parallel calculation cases. However, normally, it is difficult to maintain a Markov chain in parallel calculation. To guarantee Markov's qualifications, the following ideas are proposed.
(5) By using a nature that at most one of K moves is accepted, at low temperature where the acceptance rate $\chi(T)$ is less than $1/K$, evaluations of moves are assigned into K processors, and are calculated simultaneously until one move is accepted. In high temperature, in case $\chi(T) > 1/K$, each processor is allowed to evaluate only one move, and one of the accepted moves is chosen at random, and the information is updated with the new data. Roussel-Ragot [23] claims that this approach guarantees Markov's qualifications.
(6) Witte et al.[25] has suggested another approach. They divided the calculation into two parts. One is that of accepted case, and the other is that of rejected case. These two parts are calculated in different processors. This approach can be extended to several hierarchies. They also suggest that the processors can be assigned to calculations of accepted case and of rejected case as unbalanced tree type manner according to the number of accepted moves. That is, if number of rejected case increases in low temperature, number of processors assigned to the rejected cases is increased, and number of processors assigned to accepted cases is decreased. By Witte[25], if hierarchical depth is three, calculation speed is around a half of the series calculation case.

2.3 Power System Applications

2.3.1 INTRODUCTION

Simulated Annealing has been applied to solve many kinds of power system's optimization problem for these 10 years. Especially it is applied to such combinatorial optimization problems as unit commitment, generator maintenance scheduling, VAR resource planning, network planning and distribution system planning (operation), etc. Table 1 summarizes power system applications of SA around these 10 years.

Table 1 Power System Applications of SA

Method Application	Simple SA	Parallel SA	Combination with other method
Generator Scheduling & Unit Commitment	Mantawy[18] Zhuang[28]	Annakkage[1]	Wong[26] Wong[27]
Generator Maintenance Scheduling	Satoh[24])		Kim[16]
VAR Planning	Chen[9] Hsiao[13] Hsiao[14]		
Distribution Systems Planning & Operation	Billinton[4] Chiang[5] Chiang[6] Chiang[7] Chiang[8] Chu[10] Jiang[15]		
Network Planning	Romero[22]	Gallego[11] Mori[19]	
System Measurement	Baldwin[2]		
Load Forecasting	Mori[20]		

In Table 1, related references are shown in the appropriate positions of the matrix where applications and noteworthy methods are shown at vertical and horizontal elements respectively. The references are selected from IEEE PES (Power Engineering Society) transactions from January 1990 to March 1998. Most applications are based on the simple SA explained in the previous section. Many applications are found in the area of distribution planning and operation. In this area, since optimal load transfer operation problem is inevitably a huge combinatorial optimization problem, it has been difficult to find an effective and accurate solution algorithm. In solving such complex combinatorial optimization problems, if computation time is not a problem, SA can be successfully applicable. Other important application areas or combinatorial optimization problems are generation scheduling (unit commitment), generator maintenance scheduling and VAR facility planning as can be seen in Table 2. Let us look how SA can be applied to these areas in the following sections.

2.3.2 APPLICATIONS TO GENERATION SCHEDULING

First paper, which applied SA to power system, is published by Zhuang,F et al. [28] unit commitment area. The unit commitment problem is defined to minimize a sum of fuel cost and start-up cost of generators under the constraints of demand/supply balance, spinning reserve, power generation limit, minimum up/down time, must-run constraint and crew constraint, etc. Zhuang has formulated the problem for a power system with I thermal units to be scheduled over an H-hour planning horizon with hourly time intervals as follows:

[Objective function]

$$\underset{u_i^t,\ p_i^t}{\text{Minimize}} \sum_{t=1}^{H}\sum_{i=1}^{I}\left[u_i^t FC_i^t(p_i^t) + SC_i(s_i^t)\right] \qquad (11)$$

where,

u_i^t, p_i^t: commitment and real power generation variables of unit i in hour t, $FC_i^t(.)$: fuel cost function of unit i in hour t, $SC_i(s_i^t)$: start-up cost of unit i in hour t incurred only when the state (the cumulative on- or off- time) s_i^t of unit i in hour t indicates a start-up transition.

[Constraints]

* Power balance constrains:

$$\sum_{i=1}^{I} u_i^t p_i^t = D^t \qquad t=1,\ldots\ldots..H \qquad (12)$$

where, D^t: system power demand in hour t.

* Spinning reserve constraints:

$$\sum_{i=1}^{I} u_i^t \overline{p}_i - D^t \geq R^t \qquad t=1,\ldots\ldots..H \qquad (13)$$

where, $\overline{p}_i$: upper generation limit of unit i, R^t: system capacity reserve requirement in hour t.

* Unit-wise constrains including:

- power generation limits, $\underline{p}_i \leq p_i^t \leq \overline{p}_i$,
- integrality constraints, $u_i^t \in (0,1)$
- minimum up/down time constraints
- state transition equation

To solve the above problem by SA, we must define a cooling schedule and must determine how to create neighborhood solutions. For this purpose, first, Zhaung has partitioned the constraints into two parts:
(1) "easy" constraints = (generation limit, minimum up- and down-time constraints and must run/unavailability constraints)
(2) "difficult" constraints = (power balance and reserve constraints and crew and short-term maintenance constraints)

The easy constraints above are satisfied when neighborhood solution is created, and the violations of difficult constraints are added to the objective function. That is,

$$\text{Minimize } c(x) + \sum_{V_j(x) \in V(x)} \pi_j V_j(x), \quad x \in X \qquad (14)$$

$X = \{x \mid x \text{ satisfies all constraints in easy constraints}\}$, $V(x)$: set of violations of difficult constraints. π_j: penalty factor, $c(x)$: original objective function.

This is solved under the easy constraints. The neighborhood of the current solution is created through the following algorithm.
(1) Pick an $i \in (1,2,..,r)$ and a $t \in (1,..,H)$ by two independent random number.
(2) Change the commitment of unit i in hour t in the current iterate. Eliminate all resulting violations on easy constraints while making as few changes on the commitment of unit i in other hours as possible (details are omitted).

For a cooling schedule, constant factor ρ is employed. That is, $T = \rho T_0$ is used. Mantawy et al[18] used a polynomial-type cooling schedule and they exploited a neighborhood solution generating algorithm in which no infeasible solution is created. Annakkage et al.[1] employed a parallel SA in solving the same problem. They implemented a parallel algorithm of type (6) shown in section 2.2.3.

In solving the above unit commitment problem, economic dispatch or short term optimal generation scheduling must be solved although simplified priority method is sometimes employed. Wong,K.P [26] proposed a solution algorithm of generation scheduling by using SA and hybrid GA(genetic algorithm)/SA method under the assumption that unit commitment has determined. In the algorithm fuel cost is minimized under the constraints of power balances, generation limit of each generator, and take-or-pay contract of multiple fuels. In solving the problem by SA, neighborhood solution is generated by perturbing generator's output and fuel allocation factor to each generator randomly. In reference [27], calculation results of SA, basic genetic algorithm (BGA), improved GA (IGA), hybrid GA/SA (GAA) and improved hybrid GA/SA (GAA2) are compared, and they concluded that the result of GAA2 is the most reliable. In GAA, acceptance after the genetic operation is determined by Boltzmann's probability (eq.(10)). In GAA2, to reduce memory size and computation burden, population size of GA is limited to two. To compensate smaller population size, the fittest chromosome generated so far is stored and re-introduced into the population in higher probability than BGA.

2.3.3 APPLICATIONS TO GENERATOR MAINTENANCE SCHEDULING

The generator maintenance scheduling problem is to determine maintenance schedule of thermal generating unit with minimizing fuel cost, maintenance cost and/or variance of spinning reserve rate. The problem is formulated as a mixed integer programming problem as follows.

[Objective function]

The objective is to minimize the sum of the following two terms:

$$\text{Minimize} \quad \sum_{i=1}^{I}\sum_{j=1}^{J} f_i(p_{ij}) + \sum_{i=1}^{I} c_i(x_i) \quad (15)$$

where, the first terms is the production cost and the second is the maintenance cost.

[Constraints]

1) Nominal starting period of maintenance of each unit:

$$x_i \in X_i \subseteq \{1,2,....,J\} \quad (16)$$

2) Continuous maintenance periods:

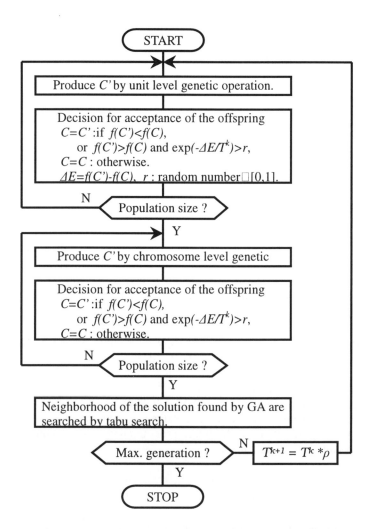

Fig. 4 Solution algorithm of generator maintenance scheduling by (GA+SA+TS)

$$y_{ij} = \begin{cases} 0, j=1,2,\ldots,x_i-1 \\ 1, j=1,2,\ldots,x_i+M_i-1 \\ 0, j=x_i+M_i,\ldots,J \end{cases} \quad (17)$$

3) Crew constraint:

$$y_{(i1)j} + y_{(i2)j} \leq 1, \quad j=1,2,\ldots,J \quad (18)$$

4) Unit maintenance sequence (order constraint):

$$x_{i1} + M_{i1} \leq x_{i2} \quad (19)$$

5) Demand constraint:

$$\sum_{i=1}^{I} p_{ij} * (1 - y_{ij}) \geq D_j + R_j, \qquad j = 1, 2, \ldots, J \qquad (20)$$

Where,
i: index of generating unit, I: number of generating units, j: index of period, J: number of planning horizon in weeks, x_i: maintenance start period $x_i \in \{1, 2, \ldots, J\}$, X_i: set of preferred maintenance start periods, M_i: maintenance length in weeks, Yij:state variable;

$$Y_{ij} = \begin{cases} 1; \text{if unit } i \text{ is in maintenance at preiod } j \\ 0; \text{otherwise} \end{cases}$$

p_{ij}: generator output of unit i at period j, f_i: fuel cost function, $c_i(x_i)$: maintenance cost of unit i when the maintenance is committed at period xi, p_i: capacity of unit i, D_j: anticipated demand at period j, R_j: required reserve at period j, k: iteration count, Δ : change of cost, r: random number uniformly distributed over an interval [0,1], T: temperature, ρ: cooling rate.

For the above problem, Satoh et al. [24] has applied a simple simulated annealing method. They start from a randomly selected initial solution (maintenance schedule). By creating a neighborhood through randomly selecting a generator unit and its maintenance period, annealing procedure is continued. Calculation results by changing a cooling rate from 0.95 to 0.98 are shown in the paper. As a general tendency, better result can be obtained if cooling rate is slow. By comparing with the results through integer programming, better solution can be found by SA with far less computation time.

Kim et al. [16] applied a combined method of simulated annealing, genetic algorithm (GA) and tabu search (TS) to more complex maintenance scheduling problem for multiple years. They took three maintenance classes (A:detailed, B:middle, C:simple) into consideration, and minimizing the variance of spinning reserve rate as well as fuel and maintenance cost. For such a complex problem, if problem size is large, SA takes too much computation time. To reduce computation time, they applied a GA based method combined with SA and TS. A flowchart of the solution algorithm is shown in Fig. 4. In the solution algorithm, the maintenance schedules are represented by fixed length of binary string as shown in Fig. 5. For such a string, two kinds of genetic operations are applied: genetic operations between sub-chromosomes (maintenance schedule of generator unit) and between chromosomes (maintenance schedule of whole generators for multiple years). After both genetic operations, new chromosomes are accepted by using SA's acceptance probability. This accelerates solution to converge to local minimum. To avoid converging to local minimum, TS is employed after new chromosome (solution) is accepted. Kim et al.[16] claims that (GA+SA+TS) can obtain the best solution if we compare the calculation results among simple GA, simple SA, (GA+SA) and (GA+SA+TS) methods.

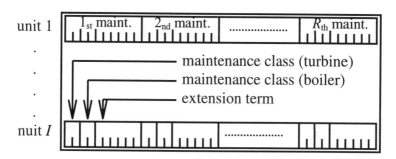

Fig. 5 String structure of (GA+SA+TS) method in Kim et al. [16]

2.3.4 APPLICATIONS TO VAR PLANNING

Normally, VAR planning problem is formulated as follows.

[Objective function]

$$\text{Minimize } F = Cost + \gamma w \sum_{l \in NC} d^l Ploss^l \qquad (21)$$

Where,

$$Cost = \sum_{i \in NB} \left(\alpha_i^{SC} SC_i^{new} + k_i^{SC} \beta_i^{SC} + \alpha_i^{ShR} ShR_i^{new} + k_i^{ShR} \beta_i^{ShR} \right) \qquad (22)$$

$$k_i^{SC} = \begin{cases} 0, (SC_i^{exist} > 0) \\ 1, (SC_i^{new} > 0, SC_i^{exist} = 0) \end{cases}, \quad k_i^{ShR} = \begin{cases} 0, (ShR_i^{exist} > 0) \\ 1, (ShR_i^{new} > 0, ShR_i^{exist} = 0) \end{cases} \qquad (23)$$

$$Ploss^l = \sum_{i \in NB} \sum_{\substack{j \in NB \\ j \neq i}} G_{ij}^l \left(V_i^{l\,2} + V_j^{l\,2} - 2V_i^l V_j^l \cos\theta_{ij}^l \right) \qquad (24)$$

[Constraints]

$$0 \le SC_i^{new} \le SC_i^{upper} \quad (i \in NB) \qquad (25)$$

$$0 \le ShR_i^{new} \le ShR_i^{upper} \quad (i \in NB) \qquad (26)$$

$$0 \le SC_i^l \le SC_i^{exist} + SC_i^{new} \quad (i \in NB, l \in NC) \qquad (27)$$

$$0 \le ShR_i^l \le ShR_i^{exist} + ShR_i^{new} \quad (i \in NB, l \in NC) \qquad (28)$$

$$\underline{T}_a \leq T_a^l \leq \overline{T}_a \qquad (a \in Br, l \in NC) \qquad (29)$$

$$\varphi_h^l(\mathbf{V}^l, \theta^l) = 0 \qquad (h \in 2NB, l \in NC) \qquad (30)$$

$$\underline{P}_{Gk} \leq P_{Gk}^l \leq \overline{P}_{Gk} \qquad (k \in NG, l \in NC) \qquad (31)$$

$$\underline{Q}_{Gk} \leq Q_{Gk}^l \leq \overline{Q}_{Gk} \qquad (k \in NG, l \in NC) \qquad (32)$$

$$\underline{V}_{Gk} \leq V_{Gk}^l \leq \overline{V}_{Gk} \qquad (k \in NG, l \in NC) \qquad (33)$$

$$\underline{V}_b \leq V_b^l \leq \overline{V}_b \qquad (b \in NL, l \in NC) \qquad (34)$$

where,

$\alpha_i^{SC}, \alpha_i^{ShR}$: variable cost of capacitor and inductor banks, $\beta_i^{SC}, \beta_i^{ShR}$: fixed cost for capacitor and inductor bank installation, SC_i^{new}, ShR_i^{new} : number of capacitor and inductor banks installed in i-th bus (discrete variables), SC_i^l, ShR_i^l : number of capacitor and inductor banks which are utilized at i-th bus in l-th load level (discrete variables), $SC_i^{exist}, ShR_i^{exist}$: number of existing capacitor and inductor banks at i-th bus, $SC_i^{upper}, ShR_i^{upper}$: upper installation limit of capacitor and inductor banks at i-th bus, T_a^l : a-th transformer's tap position in l-th load level (discrete variable), V^l, θ^l : voltage magnitude and angle in l-th load level (continuous variable vectors), φ_h^l : h-th equation of power flow equations in l-th load level, P_{Gk}^l, Q_{Gk}^l : active and reactive power at k-th generator in l-th load level, V_{Gk}^l : voltage magnitude at k-th generator bus in l-th load level, G_{ij}^l : admittance of transmission line between i-th bus and j-th bus, γ: unit energy loss cost, w: weighting factor, d^l : duration time of l-th load level, NC: set of load level numbers, NB: set of bus numbers, NL: set of load bus numbers, NG: set of generator bus numbers, NT: set of transmission line numbers, Br: set of transformer numbers, $\underline{\bullet}, \overline{\bullet}$: lower and upper limit

In the above formulation, several discrete variables exist, and the problem appears to be a combinatorial optimization problem. Hsiao et al.[13] proposes a solution algorithm by SA for the above problem. To generate neighborhood solutions, they try to perturb VAR resource configuration, and calculate modified fast de-coupled load flow for several contingency cases to check the feasibility. After satisfactory perturbations and temperature reduction, for each node, the maximum number of necessary VAR resources among contingencies is selected as installation size.

They suggest that cooling rate for the problem is 0.85-0.95, and number of moves in one temperature is given by $10\ln(2^n)$ where n is the number of total buses. Final temperature is found when sampled mean value of cost function does not change or the acceptance rate is less than 1%. This solution algorithm is extended to solve multi-objective problem through two stages SA in Hsiao et al. [14]. Since two stage SA by the same authors will be also found in the next section for distribution planning applications, explanation is disregarded here.

2.3.5 APPLICATIONS TO DISTRIBUTION SYSTEMS OPTIMIZATION

Many kinds of combinatorial optimization problems exist in distribution systems planning and operation area. SA has been applied to mainly two regions of the distribution planning area: one is facility installation planning and the other is system reconfiguration or network planning. Billington[4] proposed a method to find where to install distribution switches so as to create the most reliable radial distribution system configuration. Chu[10], Chiang [7], Chiang [8], and Jiang [15] proposed SA applications to capacitor installation problem and capacitor bank operational problem in a distribution system. Since capacitor installation planning applications are discussed in the previous section, let us look into a distribution network reconfiguration problem in this section.

Chiang et al. [7] has applied SA to distribution system's reconfiguration problem. They formulated the problem as a multi-objective combinatorial optimization problem as shown below.

[Objective function]

$$\min_{x_1} c_1(x_1, x_2)$$
$$\min_{x_1} c_2(x_1, x_2) \tag{35}$$

[Constraints]

$$F(x_1, x_2) = 0$$
$$G(x_1, x_2) \leq 0 \tag{36}$$

Where, c_1 and c_2 mean distribution loss and feeder power balance respectively. x_1 represents discrete variables to show switch status and x_2 represents continuous variables to calculate power flow of distribution system. To solve the above multi-objective problem by SA, they proposed the following solution algorithm.

[Solution algorithm]

(Step 1) Apply a perturbation mechanism $x_1^i \longrightarrow x_1^{i+1}$.

(Step 2) Compute x_2^{i+1} from $F(x_1^{i+1}, x_2) = 0$.

(Step 3) Check the feasibility
- If $G(x_1^{i+1}, x_2^{i+1}) \leq 0$, go to (Step 4);
- Otherwise go to (Step 1).

(Step 4) Set $x_1^i = x_1^{i+1}$ if one of the following conditions is satisfied
- $\Delta c_1(x^i) \leq 0$ and $\Delta c_2(x^i) \leq 0$;
- $\min\left[e^{-\frac{\Delta c_1(x^i)}{T}}, e^{-\frac{\Delta c_2(x^i)}{T}}\right] \geq \text{random}[0,1)$

Otherwise, leave it unchanged.
Go to (Step 1).

From the numerical results in Chiang *et al[5]* they claims that both voltage profile and loss minimization are satisfactory attained by applying the method.

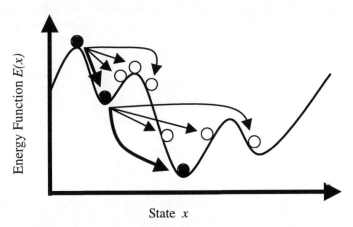

Fig. 6 Search of solution space by parallel SA in Mori *et al.* [19]

2.3.6 APPLICATIONS TO NETWORK PLANNING

SA is also applied to transmission network planning problem. Mori *et al.* [19] applied parallel SA to power system network decomposition problem. They search solution space by calculating several deviations simultaneously from the current solution, and move to the solution with minimum objective function among deviated solutions, as shown in Fig. 6. They claim that possibility to find the global optimal can be increased due to multi-point search through their parallel SA.

Gallego *et al.* [11] have proposed a sequential SA and parallel SA algorithms to solve the following long term transmission network expansion planning problem.

[Objective function]

$$\text{Minimize} \quad v = \sum_{ij} c_{ij} n_{ij} + \sum_{i} \alpha_i r_i \tag{37}$$

[Constraints]

$$B(x+\gamma^o)\theta + g + r = d \tag{38}$$

$$(x_{ij} + \gamma_{ij}^o)|\theta_i - \theta_j| \le (x_{ij} + \gamma_{ij}^o)\overline{\phi}_{ij} \tag{39}$$

$$0 \le g \le \overline{g} \tag{40}$$

$$0 \le r \le d \tag{41}$$

$$0 \le n_{ij} \le \overline{n}_{ij} \tag{42}$$

where,
c_{ij}: cost of the addition of a circuit in branch i-j, $B(.)$: susceptance matrix, θ: vector of nodal voltage angles, γ^o: vector of initial susceptances, whose elements are γ_{ij}^o: i.e. sum of the susceptances in branch i-j at the beginning of the optimization, n_{ij}: number of circuits added in branch i-j, $n_{ij} = x_{ij} / \gamma_{ij}$; where γ_{ij} is the susceptance of the new circuits, $\overline{\phi}_{ij}$: ratio $\overline{\phi}_{ij} = \overline{f}_{ij}/\gamma_{ij}$; where $\overline{f}_{ij}$ is the maximum flow in a circuit i-j, d: vector of liquid demand, g: generation vector, $\overline{g}$: vector of maximum generation capacity, r: vector of artificial generations, α: penalty parameter associated with loss of load caused by lack of transmission capacity.

The implemented sequential simulated annealing method is the ordinal one as shown below. To compare the algorithm with the parallel one, implementation details are also added from Gallego et al. [11]

Sequential Algorithm:
(Step 1) Initialize current solution x_o, temperature T_o, $k = 0$ and make $x_{opt} = x_j$.

(Step 2) Evaluate cost function $v(x_j)$;
  If $v(x_j) < v(x_{opt})$, update incumbent solution $x_{opt} = x_j$
(Step 3) Determine state transition Δx_j.
(Step 4) Evaluate the cost of the candidate solution, $v(x_j+\Delta x_j)$,
  and determine the cost variation, $\Delta v = v(x_j+\Delta x_j)-v(x_j)$.
(Step 5) If $\Delta v < 0$, go to (Step 7).
(Step 6) If $\exp(-\Delta v/T_k) < R$, where R is a random number in the range [0,1],
  go to (Step 7); otherwise, go to (Step 8).
(Step 7) Update the current state, $x = x + \Delta x$;
  Test whether the system reached thermal equilibrium;
  if so go to (Step 8), otherwise go to (Step 2).
(Step 8) Reduce temperature and make $k = k + 1$.
(Step 9) If $T_k < T_{min}$, stop; otherwise, go to (Step 2).

Implementation Details:
At (Step 7) of the Sequential Algorithm (given above) a certain number, N_k, of trials is performed at each temperature level until a near thermal equilibrium condition is reached. In practice, rather than actually testing whether thermal equilibrium has been reached, it is easier to specify a limit for N_k, i.e. the number of trials which are performed at temperature level T_k. Normally, this is done by specifying the number of trials N_o for the initial temperature and a control parameter ρ; the number of trials increases with the temperature level according to $N_{k+1} = \rho N_k$ ($\rho = 1.2$ is used).

Gallego et al. [11] also provides a parallel SA algorithm to solve the above expansion planning problem. In the parallel algorithm, to maintain the Markov's qualifications, the following strategy is employed.

For each temperature level, a set of Nk trials is broken down into a number of sub-set equal to the number of processors of the parallel machine, each one performing Nk/np trials. Where, np is the number of processors in the parallel machine.

The parallel algorithm employed by them is shown below.

Parallel Algorithm:
(Step 1) If $p = 0$: (a) initialize x_o and T_o; (b) broadcast x_o.
  If $p = 0,\ldots,np-1$:(a) receive x_o; and (b) make $x_o^p = x_o$ and $x_{opt}^p = x_o$.

(Step 2) Evaluate cost function $v(x_j^p)$;

  If $v(x_j^p) < v(x_{opt}^p)$, update incumbent solution at node p, $x_{opt}^p = x_j^p$.

(Step 3) Determine state transition at node $p, \Delta x_j^p$.

(Step 4) Evaluate the cost of the candidate solution, $v(x_j^p + \Delta x_j^p)$,

and determine the cost variation, $\Delta v^p = v(x_j^p + \Delta x_j^p) - v(x_j^p)$.

(Step 5) If $\Delta v^p < 0$, go to (Step 7).

(Step 6) If $\exp(-\Delta v^p/T_j) < R^p$, where R^p is a random number in the range [0,1], go to (Step 7); otherwise, go to (Step 8).

(Step 7) Update the current state, $x^p = x^p + \Delta x^p$;
test thermal equilibrium conditions at node p;
if so go to (Step 8); otherwise go to (Step 2).

(Step 8) If $p = 1,\ldots,np$-1: (a) send x_{opt}^p to node 0; and

(b) receive new global x_{opt} from node 0; (c) make $x_{opt}^p = x_o$; (d) go to (Step 2).

If $p = 0$: (a) update temperature $T_{k+1} = \beta T_k$; (b) make $k = k+1$;

(c) receive x_{opt}^p form nodes $p = 1,\ldots,np$-1;

(d) find $x_{opt} = \min[x_{opt}^p]$, for $p = 0,\ldots,np$-1; (e) broadcasts x_{opt}; and

(f) go to (Step 9).

(Step 9) If $T_k < T_{min}$, stop; otherwise, go to (Step 2).

Implementation Details:

The same comments made above regarding near thermal equilibrium conditions hold for the parallel version of the simulated annealing algorithm. At (Step 8), when node 0 determines x_{opt}, it also checks whether optimal solutions coming from all nodes, $p = 1,\ldots,np$-1, happen to be the same; if so, the process stops and that solution is called the optimal one. This condition normally is satisfied before the minimum temperature T_{min} is reached, which guarantees faster convergence when compared with the Sequential Algorithm. These two conditions combined are responsible for speedups which for some runs may be much higher than the ones we could expect from the simple division of the effort of generating a Markov chain over the parallel machine processors.

They applied the both (sequential and parallel) algorithms to 45 nodes Southern Brazilian System. The numerical calculations show that the parallel algorithm can find a better solution than the sequential one. It also shows that, for 45 nodes system, number of 16 or 32 parallel processors are necessary not to reduce the computational speed but to obtain an accurate solution.

2.3.7 OTHER APPLICATIONS

Several other applications of the simulated annealing method have been proposed in solving optimization problems in a power system area. For example, Baldwin [2] applied the SA based search algorithm to the minimal phasor measurement unit placement problem. Mori [19] applied SA to short-term load forecasting problem.

Generally speaking, SA can be effectively applied to combinatorial optimization problems because it is difficult to develop an efficient mathematical programming algorithm for solving them. From experiences of many applications, we are sure that SA can find quiet near optimal solutions for such complex problems as combinatorial, non-linear, etc. although much computational time is necessary to obtain a satisfactory accurate solution. More applications of SA can be expected in new areas where SA has not yet been applied.

2.4 Concluding Remarks

Since SA can find a near optimal solution in most of the experienced application cases, SA can be used as a reference algorithm to evaluate a newly developed algorithm. That is, the solution accuracy can be evaluated by comparing the calculation results of the developed algorithm with those of SA in which starting temperature and cooling schedules are carefully determined. Also, we can use the nature that the solution of SA is accurate enough when starting temperature is high and cooling rate is satisfactory slow. Namely, though computational speed is slow, we can estimate the optimal solution within finite time by applying SA to the problem for which the optimal solution is not yet found because of huge computation burden of an existing method. An algorithm of simulated annealing method is so simple that every engineer in power area can code it for any optimization problem, and the algorithm can be easily applied for the above purposes in a power area.

2.5 References

1. Annakkage, U.D. , Numnonda,T. and Pahalawaththa,N.C. (1995) Unit commitment by parallel simulated annealing, *IEE Proc.-Gener. Trans. Distrib.*, **6**, 595-600
2. Baldwin, T. *et al.* (1993) Power system observability with minimal phasor measurement placement, *IEEE transactions on Power Systems*, **8**, 707-715
3. Beckerman, M. (1997) *Adaptive Cooperative Systems*, John Wiley & Sons, Inc.
4. Billinton, R. and Jonnavithula, S. (1996) Optimal switching device placement inradial distribution system, *IEEE transactions on Power delivery*, **11**, 1646-1651
5. Chiang, H.D. and Jean-Jumeau,R. (1990) Optimal network reconfigurations in distribution systems: part 1: a new formulation and a solution methodology, *IEEE transactions on Power delivery*, **5**, 1902-1908
6. Chiang, H.D. and Jean-Jumeau,R. (1990) Optimal network reconfigurations in distribution systems: part 2: solution algorithm and numerical results, *IEEE transactions on Power delivery*, **5**, 1568-1574
7. Chiang, H.D. *et al* (1995) Optimal capacitor placement, replacement and control in large-scale unbalanced distribution systems: system modeling and a

new formulation, *IEEE transactions on Power Systems*, **10**, 356-362
8. Chiang, H.D. *et al* (1995) Optimal capacitor placement, replacement and control in large-scale unbalanced distribution systems: system solution algorithms and numerical studies, *IEEE transactions on Power Systems*, **10**, 363-369
9. Chen, Y.L. and Liu, C.C. (1995) Optimal multi-objective VAR planning using an interactive satisfying method, *IEEE transactions on Power Systems*, **10**, 664-669
10. Chu, R.F., Wang, J.C. and Chiang, H.D (1994) Strategic planning of LC compensators in nonsinusoidal distribution systems, *IEEE transactions on Power delivery*, **9**, 1558-1563
11. Gallego, R.A., Alves, A.B., Monticelli, A. and Romero, R. (1997) Parallel simulated annealing applied to long term transmission network expansion planning, *IEEE transactions on Power systems*, **12**, 181-188
12. Geman, S. and Geman, D. (1984) Stochastic relation, Gibbs distribution, and Bayesian restoration of images, *IEEE Transactions on Pattern Analysis and machine Intelligence* **PAMI-6**, 721-741
13. Hsiao, Y.T. *et al* (1993) A new approach for optimal VAR sources planning in large scale electric power systems, *IEEE transactions on Power Systems*, **8**, 988-996
14. Hsiao, Y.T. *et al* (1994) A computer package for optimal multi-objective VAR planning in large scale power systems, *IEEE transactions on Power Systems*, **9**, 668-676
15. Jiang, D. and Baldick, R. (1996) Optimal electric distribution system switch reconfiguration and capacitor control, *IEEE transactions on Power Systems*, **11**, 890-897
16. Kim, H. , Hayashi, Y. and Nara, K. (1997) An algorithm for thermal unit maintenance scheduling through combined use of GA SA and TS, *IEEE transactions on Power Systems*, **12**, 329-335
17. Kirkpatrick, S., Gelatt C.D. and Vecchi, M.P. (1983) Optimization by simulated annealing, *Science*, **220**, 671-680
18. Mantawy, A.H. , Abdel-Magid, Y.L. and Selim,S.Z. (1998) A simulated annealing algorithm for unit commitment, *IEEE transactions on Power Systems*, **13**, 197-204
19. Mori,H. and Takeda, K. (1994) Parallel simulated annealing for power system decomposition, *IEEE transactions on Power Systems*, **9**, 789-795
20. Mori, H. and Kobayashi, H (1996) Optimal fuzzy inference for short-term load forecasting, *IEEE transactions on Power Systems*, **11**, 390-396
21. Reeves, C.R. (1993) *Modern Heuristic Techniques for Combinatorial Problems*,
22. Romero, R., Gallego, R.A. and Monticelli,A. (1996) Transmission system expansion planning by simulated annealing, *IEEE transactions on Power Systems*, **11**, 364-369
23. Roussel-Ragot, P. and Dreyfus, G. (1990) A problem independent parallel

implementation of simulated annealing: models and experiments, *IEEE Transactions on Computer-aided Design*, **9**, 827-835
24. Satoh, T. and Nara,K. (1991) Maintenance scheduling by using simulated annealing method, *IEEE transactions on Power Systems*, **6**, 850-857
25. Witte, E.E. and Chamberlain,R.D. (1991) Parallel simulated annealing using speculative computation, *IEEE transactions on Parallel and Distributed Systems*, **2**, 483-493
26. Wong, K.P. and Wong,S.Y.W. (1996) Combined Genetic algorithm / simulated annealing / fuzzy set approach to short-term generation scheduling with take-or-pay fuel contract, *IEEE transactions on Power Systems*, **11**, 128-136
27. Wong, K.P. and Wong,S.Y.W. (1997) Hybrid genetic / simulated annealing approach to short-term multiple fuel-constrained generation scheduling, *IEEE transactions on Power systems*, **12**, 776-784
28. Zhuang, F. and Galiana,F.D. (1990) Unit commitment by simulated annealing, *IEEE transactions on Power Systems*, **5**, 311-318

Chapter 3

TABU SEARCH APPLICATION IN FAULT SECTION ESTIMATION AND STATE IDENTIFICATION OF UNOBSERVED PROTECTIVE RELAYS IN POWER SYSTEM

Fushuan Wen
Dept. of Electrical Engineering
Zhejiang University
Hangzhou, 310027
Zhejiang Province
P.R.China

C.S. Chang
Dept. of Electrical Engineering
National University of Singapore
10 Kent Ridge Crescent
Singapore 119260
Republic of Singapore

3.1 Introduction

Fault section estimation[1] aims at identifying faulty components (sections) in a power system by using information on operations of protective relays and circuit breakers. Several kinds of methods have so far been developed, such as the logic-based[2,3], expert system-based[4-6], artificial neural network-based[7] and optimization-based[8-14] methods. Of these methods, the expert system-based method is the most established. Up to now, many kinds of expert systems have been developed using the rule based[4] and model based[5,6] methods. In order to achieve precise inference especially for the complex fault cases, the rule based expert system must involve a great number of rules describing the complex protection system behaviour. Maintenance of a large knowledge base is very difficult. On the other hand, the model based system is easy to maintain, but the inference process is time consuming.

In recent years, the application of Artificial Neural Networks (ANNs) to the fault section estimation has been an active research area. This method treats the fault section estimation as a classification problem, and uses the appropriate ANNs such as the back-propagation model or the Kohonen model to train and estimate[7]. It is difficult to reasonably specify a sample set, so the correctness of the result can not be guaranteed theoretically.

Recently, a kind of new methods to the fault section estimation are developed using optimization techniques[8-14]. The principle behind these methods is to formulate the fault section estimation as an optimization problem and use a global optimization method such as Boltzmann machine[8], Genetic Algorithm[9-13] (GA) or Tabu Search[14] (TS) to solve it. The methods presented in [8,9] utilize the information from protective relays, but this information is not complete in many electric dispatching centers. The method developed in [10] uses the time sequence information of circuit breakers only, which is not easy to obtain. Further development to the work presented in [9,10] is reported in [11-14]. In [11], a new 0-1 integer programming model for the fault section estimation is developed, which utilizes not only the operational information of protective relays, but also the tripping information of circuit breakers. The method presented in [11] is further improved and extended to solve the fault diagnosis problem in distribution systems in [12]. A probabilistic approach is proposed to account for the reliabilities of protective relays and circuit breakers in [13]. A protection system simulator based method is presented in [14], which also utilizes the information both from protective relays and circuit breakers. Although the work presented in [8-14] is preliminary, simulation results have shown that this kind of methods are of great promise for large scale power systems.

Most of the fault section estimation methods proposed so far are based on the premise that the status information is available from every protective relay in the given power system. In practice, due to the cost of wiring the protective relays to the remote terminal units (RTU) and upgrading the communication links to carry these additional data, only a subset of this information may be available at the Energy Management System (EMS)[15]. As pointed out in [15], ignoring this fact will force the diagnosis to become either brittle (no solution will be found) or undiscriminating (too many solutions will be offered without ranking). An incremental approach, which allows the utility to improve the performance of the diagnosis system as more protective relay information becomes available, is thus preferable[15]. Recently, some attempts have been made to solve the fault section estimation under incomplete information from protective relays, such as the work presented in [11]. In [11], the essential idea to deal with the incomplete information from protective relays is to make full use of the state information from circuit breakers (so as to complement the incomplete information from protective relays), but this effort has only achieved a limited success. This is because the states of those protective relays whose information is not available in the dispatching centers are simply treated as "non-operated" in [11], and this is certainly not appropriate. In summary, the fault section estimation problem under incomplete information from protective relays is still not solved, and to the best of our knowledge, a formal and systematic approach to this problem is still not available.

In this chapter, a formal and systematic way for solving the fault section estimation problem under incomplete information from protective relays is described. The protective relays whose information is not available (i.e., the unobserved protective relays) are explicitly distinguished from those certainly non-operated protective relays. A 0-1 integer programming model describing this problem is formulated. A concept of "state identification of unobserved protective relays" is introduced. The developed model leads to an integrated approach to the fault section estimation and state identification of the unobserved protective relays (FSE-SIUPR). The developed approach here takes the following three main steps. Firstly, based upon the causal relationship among section fault, protective relay operation and circuit breaker trip, the expected states of protective relays and circuit breakers under the faults identified by a diagnosis hypothesis are obtained by using the method developed in [11]. However, the premise of the method developed in [11] is that the state information is available from every protective relay and every circuit breaker, and as the results the states of the unobserved protective relays are simply treated as "non-operated" under any circumstances. In this chapter, the work presented in [11] is revised and extended so as to explicitly distinguish the unobserved protective relays from those certainly non-operated protective relays. Secondly, based upon the well developed parsimonious set covering theory[16], the problem of FSE-SIUPR is formulated as a 0-1 integer programming model. Thirdly, a tabu search (TS) method[17-20] is adopted for solving this problem. TS has emerged as a new, highly efficient, search paradigm for quickly finding high quality solutions to combinatorial optimization problems. It is characterised by gathering knowledge during the search, and subsequently profiting from this knowledge. As a heuristic search strategy for solving optimization problems, TS has achieved impressive practical successes in extensive areas[17-20]. Finally, a sample power system is served for demonstrating the correctness of the developed model and the efficiency of the TS-based method.

In addition, an efficient method[9] for the identification of faulty subnetworks which include all possible faulty sections is adopted by using a network configuration determination method. When faults occur in a power system, the protective relays corresponding to the faulty sections should operate to take the faulty sections out of operation. The faulty sections will be disconnected from the healthy part of the power system after the operation of the related protective relays and circuit breakers. Thus, some isolated subnetworks will be formed eventually which include all faulty sections and disconnect with all operating generators, and the fault section estimation can be confined to these subnetworks only. The method was tested using an actual power system which consists of 43 substations, 523 sections, 412 circuit breakers, 23 three-winding transformers and 77 transmission lines and two-winding transformers. This system has 107 buses under normal operating conditions. In this work, many complicated fault scenarios have been tested, and the simulation results show that only 1 to 2 seconds are required to identify all faulty subnetworks for each fault scenario on a 486 microcomputer. In this way, the size of the FSE-SIUPR problem can be greatly reduced, and very fast computation can be achieved even for large scale power systems.

3.2 The Overall Approach for FSE-SIUPR

The developed approach for the problem of FSE-SIUPR includes several steps:

a. To input the system data, including the topological structure of the given power system, operating states of each device, the set of the unobserved protective relays and the reported alarms (i.e., the operation of protective relays and circuit breakers).

b. To identify the faulty subnetworks using the method introduced in [9].

c. To estimate faulty sections from the above faulty subnetworks and identify the states of the unobserved protective relays which protects the components (sections) in the faulty subnetworks with the following:

(1) To formulate the 0-1 integer programming model as presented in Section 3 for problem of FSE-SIUPR.

(2) To produce some hypotheses randomly (each hypothesis identifies the assumed faulty sections and the assumed states of the unobserved protective relays). The assumed faulty sections should be some of the combinations of those in faulty subnetworks, and the assumed states of the unobserved protective relays should only include the states of those unobserved protective relays which protect the components (sections) in the faulty subnetworks. .

(3) To apply the TS, which is described in Section 4, to search for better hypotheses progressively until one or more hypotheses that can reasonably explain the reported alarms have been found.

3.3 The Mathematical Model for FSE-SIUPR

3.3.1 THE EXPECTED STATES OF PROTECTIVE RELAYS AND CIRCUIT BREAKERS

This subsection is extended from [11]. The main revision is that the protective relays whose states are not available are explicitly distinguished from those protective relays which are certainly non-operated. While in [11], the states of the unobserved protective relays are simply treated as non-operated.

The fault section estimation problem is to find the hypothesis or hypotheses which can explain the reported alarms. In other words, when the faults identified by a correct hypothesis or hypotheses occur in a given power system, the expected states of protective relays and circuit breakers should be consistent with the reported alarms as much as possible. Thus, a key problem is how to determine the expected states of protective relays and circuit breakers. Fortunately, according to the operating logics of protective relays and circuit breakers, their expected states can be obtained. For the convenience of presentation, we define several symbols first.

n is the total number of sections in a given power system.

n_r, n_{or}, n_{ur}, and n_c are the total numbers of the protective relays, the observed protective relays (i.e., the protective relays whose state information is available in the dispatching centers), the unobserved protective relays, and the circuit breakers in a given power system, respectively. $n_r = n_{or} + n_{ur}$.

S is a n-dimension vector, and its ith element, s_i, represents the ith section and its state, and $s_i = 0$ or 1 corresponds to its normal or faulty state. S is a vector to be determined.

R^+ is a n_{or}-dimension vector and denotes the actual states of the n_{or} observed protective relays. The kth element of R^+, r_k^+, represents the kth observed protective relay and its actual state, and $r_k^+ = 0$ or 1 corresponds to its non-operated or operated state, respectively.

R^0 is a n_{ur}-dimension vector and denotes the states of the n_{ur} unobserved protective relays. The kth element of R^0, r_k^0, represents the kth unobserved protective relay and its state (to be identified), and $r_k^0 = 0$ or 1 corresponds to its non-operated or operated state, respectively. R^0 is a vector to be determined.

R is a n_r-dimension vector and denotes the actual states of the n_r protective relays. The kth element of R, r_k, represents the kth protective relay and its actual state, and $r_k = 0$ or 1 corresponds to its non-operated or operated state, respectively. R is composed of R^+ and R^0. $R = (R^+, R^0)$, and this means that the observed protective relays are ranked in its first n_{or} bits, and the unobserved protective relays are ranked between bits $n_{or}+1$ and n_r. Only some elements of R, i.e., R^+, are known.

$R^*(S,R)$ is a n_r-dimension vector and denotes the expected states of the n_r protective relays. The kth element of $R^*(S,R)$, $r_k^*(S,R)$, represents the kth protective relay and its expected state. If the kth protective relay should not operate, then $r_k^*(S,R)$ should be 0, otherwise be 1. $R^*(S,R)$ is dependent on S and R.

C is a n_c-dimension vector and denotes the actual states of the n_c circuit breakers. The jth element of C, c_j, represents the jth circuit breaker and its actual state, and $c_j = 0$ or 1 corresponds to its closed (nontripped) or tripped state, respectively.

$C^*(S,R)$ is a n_c-dimension vector and denotes the expected states of the n_c circuit breakers. The jth element of $C^*(S,R)$, $c_j^*(S,R)$, represents the jth circuit breaker and its expected state. If the jth circuit breaker should not trip, then $c_j^*(S,R)$ should be 0, otherwise be 1. $C^*(S,R)$ is dependent on S and R.

Now, we use a simple example as described in the Appendix (Subsection 7.1) to illustrate how to determine $r_k^*(S,R)$ and $c_j^*(S,R)$.

According to the operating logics of main protective relays (see Subsection 7.1), it is easy to determine their expected states. For example, according to the operating logic of r_1 (i.e., Am), it is known that if a fault occurs on s_1 (i.e., A), r_1 should operate. Thus, we have:

$$r_1^*(S,R) = s_1 \qquad (1)$$

Similarly, we can get:

$$r^*_2(S,R) = s_2 \tag{2}$$
$$r^*_3(S,R) = s_3 \tag{3}$$
$$r^*_4(S,R) = s_4 \tag{4}$$
$$r^*_5(S,R) = s_4 \tag{5}$$
$$r^*_6(S,R) = s_5 \tag{6}$$
$$r^*_7(S,R) = s_5 \tag{7}$$

The expected states of backup protective relays are more difficult to determine than those of the main protective relays. For example, according to the operating logic of r_8 (L1Ap), it is known that if a fault occurs on s_4 (L1) and r_4 (L1Am) failed to operate, then r_8 should operate. Thus, we have:

$$r^*_8(S,R) = s_4(1-r_4) \tag{8}$$

Similarly, we can get:

$$r^*_9(S,R) = s_4(1-r_5) \tag{9}$$
$$r^*_{10}(S,R) = s_5(1-r_6) \tag{10}$$
$$r^*_{11}(S,R) = s_5(1-r_7) \tag{11}$$
$$r^*_{12}(S,R) = 1-[1-s_2(1-c_3)][1-s_5(1-c_3)(1-c_4)] \tag{12}$$
$$r^*_{13}(S,R) = s_1(1-c_2) \tag{13}$$
$$r^*_{14}(S,R) = s_3(1-c_5) \tag{14}$$
$$r^*_{15}(S,R) = 1-[1-s_2(1-c_4)][1-s_4(1-c_3)(1-c_4)] \tag{15}$$

The expected states of circuit breakers are more difficult to determine than those of protective relays. This is because the expected states of most circuit breakers are related to two or more protective relays. For example, CB2 is related to Am (r_1), L1Am (r_4), L1Ap (r_8) and L1As (r_{12}).

According to the operating logic of r_1 (Am), it is known that if a fault occurs on s_1 (A), then r_1 should operate to trip c_1 (CB1). Thus, we have:

$$c^*_1(S,R) = s_1 r_1 \tag{16}$$

According to the operating logics of r_1 (Am), r_4 (L1Am), r_8 (L1Ap) and r_{12} (L1As), we have:

$$c^*_2(S,R) = \text{MAX} <s_1r_1, s_4r_4, s_4(1-r_4)r_8, \{1-[1-s_2(1-c_3)][1-s_5(1-c_3)(1-c_4)]\}r_{12} \tag{17}$$

where the symbol MAX means to extract the greatest value from the set. Similarly, we can get:

$$c^*_3(S,R) = \text{MAX} <s_2r_2, s_4r_5, s_4(1-r_5)r_9, s_1(1-c_2)r_{13}> \tag{18}$$
$$c^*_4(S,R) = \text{MAX} <s_2r_2, s_5r_6, s_5(1-r_6)r_{10}, s_3(1-c_5)r_{14}> \tag{19}$$
$$c^*_5(S,R) = \text{MAX} <s_3r_3, s_5r_7, s_5(1-r_7)r_{11}, \{1-[1-s_2(1-c_4)][1-s_4(1-c_3)(1-c_4)]\}r_{15}> \tag{20}$$
$$c^*_6(S,R) = s_3 r_3 \tag{21}$$

From the above Eqn. (1) through Eqn. (21), it is known that some elements of $R^*(S,R)$ and $C^*(S,R)$ are both the functions of S and R. All elements in S, and some elements in R (i.e., all elements in R^0) are unknown and to be determined. The determinations of the elements in S and the elements in R^0 are the objectives of the fault section estimation and the state identification of the unobserved protective relays, respectively. Thus, these two subproblems can not be dealt with separately, and instead, they must be solved integratedly.

3.3.2 THE SET COVERING THEORY FOR THE PROBLEM OF FSE-SIUPR

As stated in Subsection 3.1, the fault section estimation is to find the most probable hypothesis or hypotheses that can explain the reported alarms. Thus, some criteria should be specified to reflect how well a hypothesis or hypotheses can explain the reported alarms, or in other words, how to define the word "explain".

One of the leading theories for the fault diagnosis problem in artificial intelligence (AI) community is based upon the notion of parsimoniously covering a set of reported alarms, R^+ and C. The premise of the parsimonious covering theory is that a diagnosis hypothesis must be a cover of R^+ and C in order to account for the presence of all alarms in R^+ and C [16]. On the other hand, not all covers of R^+ and C are equally plausible as hypotheses for a given problem. The principle of parsimony, or "Occam's Razor", is adopted as a criterion of plausibility: a "simple" cover is preferable to a "complex" one. Therefore, a plausible hypothesis, or an explanation of R^+ and C, is defined as a parsimonious cover of R^+ and C, i.e., a set of fault sections S_c and a set of the assumed states for the unobserved protective relays R_c^0 that covers R^+ and C and satisfies some notion of being parsimonious or "simple". Thus, an essential problem in this theory is: what is the nature of "parsimony" or "simplicity" ? or in other words, what makes one cover of R^+ and C more plausible than another? Mathematically, this problem is equivalent to how to define a suitable criterion to describe "parsimony". Up to now, several different criteria have been proposed[16], such as: (1) Single fault restriction: a cover S_c and R_c^0 of R^+ and C is an explanation if it contains only a single fault section or only one operated protective relay from the unobserved protective relays; (2) Relevancy: a cover S_c and R_c^0 of R^+ and C is an explanation if it causally associate with at least one of the alarms in R^+ and C; (3) Irredundancy: a cover S_c and R_c^0 of R^+ and C is an explanation if it has no proper subsets which also cover R^+ and C; (4) Minimality: a cover S_c and R_c^0 of R^+ and C is an explanation if it has the minimal cardinality among all covers of R^+ and C, i.e., it contains the smallest possible number of fault sections and the smallest possible number of operated protective relays from the unobserved protective relays needed to cover R^+ and C.

Obviously, the single fault restriction criterion is not appropriate for our problem, because multiple section faults may happen simultaneously and more than one protective relays from the unobserved protective relays may be operated in a fault scenario. On the other hand, the relevancy criterion is not a good one for our problem, because it is too loose and may result in many solutions. The irredundancy criterion is generally quite attractive one, but the set of all irredundant covers may itself be quite large in some applications, and may contain many explanations (solutions) of very little probability. The minimality criterion is intuitively a reasonable one, because the occurring probability of complex faults is generally smaller than that of simple faults. A modified minimality criterion is adopted in this chapter.

As stated above, a reasonable diagnosis hypothesis should satisfy the above described parsimonious criterion, but this is not the only requirement for it. A more important requirement is that the expected alarms of the faults identified by the hypothesis should not be included in the set of protective relays and circuit breakers which are certainly not operated or not tripped. Thus, a reasonable diagnosis hypothesis should satisfy the following three requirements:

a. First, the hypothesis should be a cover of R^+ and C. This is because the malfunction probabilities of protective relays and circuit breakers are generally small for most power systems.

b. Second, the expected alarms corresponding to the hypothesis should not be included in the set of protective relays and circuit breakers which are certainly not operated or not tripped.

c. Third, the hypothesis with the minimum number of fault sections and the minimum number of operated protective relays from the unobserved protective relays is preferred to more complex explanations. This requirement reflects the minimality criterion.

3.3.3 THE OBJECTIVE FUNCTION OF THE PROBLEM OF FSE-SIUPR

For the problem of FSE-SIUPR, the modified minimality criterion can be described as:

Minimize $E(S,R^0) = w_1 \times (|\nabla R| + |\nabla C|) + w_2 \times (|\Delta R| + |\Delta C|) + w_3 \times (|S| + |R^0|)$ (22)
where:

∇R is a n_{or}-dimension vector, and is determined using the following method: if the jth element ($j=1, 2,...,n_{or}$) of R^+ is 1, and the jth element of $R^*(S,R)$ is 0, then set the jth element of ∇R to be 1, otherwise to be 0. ∇R reflects how well a hypothesis denoted by S and R^0 can account for the operated protective relays. If ∇R is a zero vector, then all the operated protective relays can be caused by the faults identified by the hypothesis.

∇C is a n_c-dimension vector, and is determined using the following method: if the jth element of C is 1, and the jth element of $C^*(R,S)$ is 0, then set the jth element of ∇C to be 1, otherwise to be 0. ∇C reflects how well a hypothesis denoted by S and R^0 can account for the tripped circuit breakers. If ∇C is a zero vector, then all the tripped circuit breakers can be caused by the faults identified by the hypothesis.

ΔR is a n_{or}-dimension vector, and is determined using the following method: if the jth element (j=1,2,...,n_{or}) of R^+ is 0, and the jth element of $R^*(S,R)$ is 1, then set the jth element of ΔR to be 1, otherwise to be 0. ΔR reflects if the expected alarms (from the protective relays only) of a diagnosis hypothesis denoted by S and R^0 are all included in the set of the reported alarms (from protective relays only). If ΔR is a zero vector, then the expected alarms (from the protective relays only) are all included in the set of the reported alarms (from the protective relays only).

ΔC is a n_c-dimension vector, and is determined using the following method: if the jth element of C is 0, and the jth element of $C^*(S,R)$ is 1, then set the jth element of ΔC to be 1, otherwise to be 0. ΔC reflects if the expected alarms (from the circuit breakers only) of a diagnosis hypothesis denoted by S and R^0 are all included in the set of the reported alarms (from circuit breakers only). If ΔC is a zero vector, then the expected alarms (from the circuit breakers only) are all included in the set of the reported alarms (from the circuit breakers only).

$|\nabla R|$, $|\nabla C|$, $|\Delta R|$, $|\Delta C|$, $|S|$, and $|R^0|$ are the numbers of nonzero elements in vectors ∇R, ∇C, ΔR, ΔC, S, and R^0, respectively.

Equation (22) is a measure to evaluate how well a diagnosis hypothesis denoted by S and R^0 can explain the reported alarms. The physical meaning of each term at the right hand side of this Equation is now explained in sequence. $|\nabla R|+|\nabla C|$ is a criterion to reflect if a solution of (22) is a cover of R^+ and C. If yes, $|\nabla R|+|\nabla C|=0$, otherwise $|\nabla R|+|\nabla C|$ reflects the proximity of a solution to a cover. The smaller is $|\nabla R|+|\nabla C|$, the more proximate to a cover the solution is. $|\Delta R|+|\Delta C|$ reflects if the expected alarms corresponding to the hypothesis are included in the set of the reported alarms. The smaller is $|\Delta R|+|\Delta C|$, the more expected alarms are included in the set of the reported alarms. If $|\Delta R|+|\Delta C|=0$, then the expected alarms are all included in the set of the reported alarms. $|S|$ represents the number of fault sections, and $|R^0|$ represents the number of operated protective relays from the unobserved protective relays corresponding to a hypothesis denoted by S and R^0. The three terms at the right hand side of Eqn. (22) respectively reflects the three requirements described at the end of Subsection 3.2.

w_1 through w_3 in Eqn. (22) are positive weighting coefficients to reflect the relative importance of these three requirements. It is our opinion that the priorities of these three requirements should decline progressively, so w_1 through w_3 should be specified to decrease successively. w_1 should be big enough, and be specified as the biggest among these three coefficients to ensure that the solution or solutions of Eqn. (22) are covers of R^+ and C. In this work, w_1, w_2, and w_3 are specified to be 10000, 100 and 1, respectively, and these values are applicable for any power systems. The minimization of $E(S,R^0)$ leads to the modified minimality criterion.

It must be noted that actually not all protective relays need to be included in R and $R^*(S,R)$, not all observed protective relays need to be included in R^+, not all unobserved protective relays need to be included in R^0, and not all circuit breakers need to be included in C and $C^*(S,R)$ for a given power system. It is very rare that many sections are suffering from a fault simultaneously or almost simultaneously, and the problem of FSE-SIUPR can be confined to one or more small faulty subnetworks only which can be identified by the developed method in [9]. Thus, only the involved protective relays / circuit breakers in these subnetworks need to be included in R, R^+, R^0, and $R^*(S,R)$ / C and $C^*(S,R)$ for the problem of fault section estimation and state identification of the unobserved protective relays. Generally, this is only a very small part of protective relays and circuit breakers for large scale power systems. Moreover, the possible faulty sections are those in the faulty subnetworks only. Thus, only the sections contained in these subnetworks need to be included in S. In this way, the size of the problem can be greatly reduced for large scale power systems.

The remaining problem is how to estimate the faulty section(s) and how to identify the states of the unobserved protective relays by utilizing Eqn. (22) and the reported alarms (i.e., R^+ and C), or in other words, how to find S and R^0 which minimizes $E(S,R^0)$. Obviously, this is a 0-1 integer programming problem. In the following section, we will introduce the Tabu Search (TS) method to solve this problem. The motivation to adopt the TS for solving this problem lies in its ability for finding the optimal solution efficiently, and this has been verified in many papers[14,17-35].

3.4 Tabu Search with Application to FSE-SIUPR

3.4.1 A BRIEF INTRODUCTION TO TABU SEARCH

With its roots going back to the late 1960's and early 1970's, the tabu search was proposed in its present form a few years ago by Glover[17-20]. As a heuristic search strategy for efficiently solving combinatorial optimization problems, TS has now become an established optimization approach that is rapidly spreading to many new fields, and has achieved impressive practical successes in extensive application areas[17-35]. For example, TS has been successfully applied to obtain optimal or suboptimal solutions to problems such as the Travelling Salesman problem (TSP), timetable scheduling and network layout design[18]. Together with Simulated Annealing (SA) and Genetic Algorithm (GA), TS has been singled out by the Committee on the Next Decade of Operations Research as "extremely promising" for the future treatment of practical applications[17].

TS is a restricted neighbourhood search technique, and is an iteration algorithm. The fundamental idea of TS is the use of flexible memory of search history which thus guides the search process to surmount local optimal solutions. To describe the workings of TS, we consider a combinatorial optimization problem in the following form:

Minimize $C(X)$ (23)

where X is a vector of dimension N, and its elements are integers. $C(X)$ is the objective function (cost or penalty function), and can be linear or nonlinear. The first step of TS is to produce an initial (current) solution $X^{current}$ either randomly or using an existing (heuristic) method to the given problem. The second step is to define a set of moves that may be applied to the current solution to produce a set of trial solutions. As an example, the move can take the form of $X^{trial} = X^{current} \pm \Delta X$. Here, ΔX is a vector with the same dimension as X. Among all the trial solutions thus produced, TS seeks the one that improves most of the objective function. In certain situations, if there are no improving moves, a fact which means some local optimum exists, TS chooses the one that least degrades the objective function. The most basic components of the TS are the *moves, tabu list* and *aspiration level (criterion)*, which are introduced below.

3.4.1.1. *Moves*

The search process of TS is implemented by the *moves*. A trial solution can be created by a move. In each iteration, a specified number of moves (the neighbourhood sampling number), NS_{max}, are executed. Many kinds of moves are currently available[17-35], and the following two kinds of moves[23] are adopted for solving the 0-1 integer programming problem of Eqn. (22).

a. Single move (denoted by m_i)
$X^{trial} = X^{current} + u_i$ (i=1, 2, ..., N) (24)
where, u_i is a vector with the same dimension as X, and its ith element is 1 (if the ith element of $X^{current}$ is 0) or -1 (if the ith element of $X^{current}$ is 1) and all the other elements are zero.

b. Exchange move (denoted by m_{ij})
$X^{trial} = X^{current} + u_i - u_j$ (i, j=1, 2, ..., N, and i≠j) (25)
where, u_i is a vector with the same dimension as X, and its ith element is 1 (if the ith element of $X^{current}$ is 0) or -1 (if the ith element of $X^{current}$ is 1) and all the other elements are zero. u_j is a vector with the same dimension as X, and its jth element is 1 (if the jth element of $X^{current}$ is 1) or -1 (if the jth element of $X^{current}$ is 0) and all the other elements are zero.

3.4.1.2 *Tabu List*

In order to prevent from returning to the local optimum just visited, the reverse move that is detrimental to achieving the optimum solution must be forbidden[17-20]. This is done by storing this move in a tabu list, which stores the attributes of some moves made. The elements of the tabu list are called tabu moves. The reverse moves are restricted from regions the search already explored.

The condition for a move to be a tabu move or not can be problem specific. For example, a move may be tabu if it leads to a solution which has already been considered in the last iterations (this is called as recency or short term condition). Recency based tabu list is a short-term memory that restricts the moves recently made. A tabu status for a move is in the memory for a number of iterations, which is known as the tabu tenure. The tabu tenure can be chosen either using static rules or dynamic rules. Static rules allow a value for the tabu tenure that remains fixed throughout the search. Dynamic rules allow the value of tabu tenure to vary. The example of this two kinds of rules are as follow[18]:

Static rules	Choose tabu tenure (T_{max}) to be a constant such that $T_{max}=7$ or $T_{max}=\sqrt{N}$, where N is a measure of problem dimension
Dynamic rules	Choose T_{max} to vary (randomly or by systematic pattern) between bounds T_1 and T_2, such as $T_1=5$ and $T_2=11$, or $T_1=0.9\sqrt{N}$ and $T_2=1.1\sqrt{N}$

The indicated values such as 7 and $\sqrt{N}$ are only suggestive. They should be obtained by experiment for a particular class of problems.

Frequency based tabu list[18] provides a type of information that complements the information provided by recency-based tabu list, broadening the foundation for selecting preferred moves. Frequency based tabu list is a long-term memory structure. It records the number of moves adopted in the search. This type of memory is important to diversify the search into moves that are less frequently used if the search appears to be trapped in a local neighbourhood.

The dimension of the tabu list is called the tabu list size. The tabu list size should grow with the size of the given problem, but how to specify the optimal tabu list size is still an open problem. In addition, how to manage the tabu list such as how long (how many iterations) a move can be retained in the tabu list is also an important problem. Many methods to implement and manage the tabu list are available, and the methods used here are described below.

In this work, the tabu list is updated iteration by iteration until the maximum permitted iteration number K_{max} has been reached. At the end of each iteration, the new move is added to the tabu list, and an old move may be removed if it has been in the tabu list for T_{max} iterations. The detailed implementation is as follows.

    a. If a single move m_i is selected in the kth iteration, then the following moves will be added to the tabu list:

-m_i, m_{ji} (j=1, 2, ...N, j≠i)

    b. If an exchange move m_{ij} is selected in the kth iteration, then the following moves will be added to the tabu list:

-m_i, m_{li} (l=1, 2, ...N, l≠i)

m_j, m_{jl} (l=1, 2, ...N, l≠j)

3.4.1.3 Aspiration Criterion (level)

The aspiration criterion is introduced in TS to determine when a tabu move can be overridden. The main purpose is to enable tabu moves that could possibly lead to an optimal solution. This criterion provides an added flexibility to choose good moves by allowing a tabu move to be overridden if its aspiration level is attained. Many implementation strategies for the aspiration level are available[17-20]. Choice of the aspiration criterion depends on the specific applications, and in this work the aspiration level is defined as: if a tabu move from the current solution $X^{current}$ can reach a solution which is better than the best solution found so far, then the aspiration level for this tabu move is attained and can be overridden.

3.4.1.4 A General Tabu Search Algorithm

A general tabu search algorithm for solving the minimization problem of C(X) can be described as follows:

    a.   Initially, X_0 is produced randomly and chosen as the initial solution.
    b.   Assume that there are p moves, where moves are $M=\{m_1, m_2, \ldots, m_p\}$, then the next possible solution is $M(X_0)=\{m_1(X_0), m_2(X_0), \ldots, m_p(X_0)\}$.
    c.   The neighbourhood of the current solution is the set of feasible solutions that can be reached by applying the moves. The neighbourhood set is $N(X_0) \subseteq M(X_0)$.
    d.   In the neighbourhood set, there could be some solutions that are reached by applying tabu moves. These solutions (tabu set) are denoted as $T(X_0)$, and $T(X_0) \subseteq N(X_0)$.
    e.   Within the tabu set, some solutions might have surpassed the aspiration criteria. This set of solutions is known as the aspirant set, denoted by $A(X_0)$, and $A(X_0) \subseteq T(X_0)$.

f. The next solution is chosen from the neighbour which is either as aspirant or not tabu and for which C(X) is minimal.
g. The search process then iterates from Step b to Step f until the terminating condition is met. The terminating condition should be determined based on the characteristics of the specific problems, and a commonly used one is that the prespecified maximum permitted iteration number, K_{max}, has reached.

The procedure can be summarized as follows:

Randomly choose an initial solution X_0.
Set the best solution $X^{best}=X_0$ and iteration count k=0.
While the terminating condition is not met,
    Generate $N(X_k)$
    Identify $T(X_k)$
    Identify $A(X_k)$
    Choose the best solution X' from the set $N(X_k)-[T(X_k)-A(X_k)]$ where C(X') is the minimum.
    $X_{k+1}=X'$
    If $C(X')<C(X^{best})$, then set the best solution $X^{best}=X'$
    k=k+1
End while

3.4.2 APPLICATIONS OF TABU SEARCH IN POWER SYSTEMS

To our knowledge, the research work on the applications of tabu search for solving some power system problems was started in 1995. The successful applications of TS have been reported recently in solving some power system problems, such as the hydro-thermal scheduling[21], unit commitment[22,29], reactive power planning and optimisation[23,24], capacitor placement[25,26,30], determination of distribution tie lines[27,28], thermal unit maintenance scheduling[31], fault diagnosis[14,32], alarm processing[33] and transmission network planning[34,35].

3.4.3 A TS BASED FRAMEWORK FOR SOLVING FSE-SIUPR

The general algorithm of the developed TS based FSE-SIUPR method can be described as follows:
a. At first, define a $n+n_{ur}$ dimension solution vector $X=(S, R^0)$, and let its first n bits corresponds to the elements in S sequentially, and its last n_{ur} bits corresponding to the elements in R^0 sequentially. Then, randomly generate an initial (current) solution X^*, and set the iteration counter k=0. Set the best solution vector $X^{best} = X^*$.

b. If k is equal to a prespecified maximum permitted iteration number K_{max}, then output X^{best} as the final result and stop. Otherwise, set k=k+1, and go to the next step.
c. Select a trial solution X^{trial} from the neighborhood of X^* by applying the two kinds of moves as defined in Subsection 3.4.1.1 and compute the corresponding $E(X^{trial})$ using Eqn. (22). Repeat this process until the specified neighborhood sampling number, NS_{max}, has been reached.
d. If X^{best} is not better than the best trial solution which has the smallest objective function value, then assign this best trial solution to X^{best}. Otherwise, go to next step.
e. X^* is updated to the best trial solution which has the smallest objective function value as evaluated in step c and the corresponding move is not in the tabu list or its aspiration level is attained. Then, put the move into the tabu list, and go to step b. If the best trial solution corresponds to a tabu move and its aspiration level is not attained, then check the next best trial solution, and repeat this step.

3.5 Test Results

We use a sample power system to test the developed method. It consists of 28 sections, 84 protective relays and 40 circuit breakers. The system diagram and the operating logics of all protective relays in this system are given in the Subsection 7.2 below.

More than 100 fault scenarios have been tested. Among them, 28 fault scenarios belong to the single section fault cases and all the others belong to the multiple section fault cases. In terms of the operating logics of the protective relays, it is concluded that all the results are correct. Due to the space limitation, only the detailed results for 10 test cases are given in Table 1. In these 10 test cases, we assume that the status information from every *main* protective relay is available (i.e., observed), while the status information from every *backup* protective relay is not available (i.e., unobserved) and is to be identified. Extensive test results have shown that the developed model is correct, and the TS based method is efficient. The optimal solutions for all test cases of this system can be obtained by setting the parameters in TS as follows:

$K_{max}=60$, $NS_{max}=35$, $T_{max}=30$

The computing time for each test case of this example is about 3 seconds on a 486 microcomputer, so the proposed method is of potential for practical applications.

It should be pointed out that the developed method can be applied to power systems where the status information from all protective relays is not available in its dispatching center, or in other words, the method can utilize the state information from circuit breakers only to solve the problem of FSE-SIUPR. Of course, this is a special circumstance. Under this circumstance, the method may offer more than one diagnosis results, this is because the available information is not sufficient for the diagnosis, and can not lead to a definite conclusion. Another special circumstance is that the status information from every protective relay and every circuit breaker is available, or in other words, R^0 is an empty set, and $R=R^+$. Under this circumstance, the method can offer definite diagnosis results. Thus, with more available information from protective relays, more definite diagnosis results can be obtained by the method. This means that the developed method is an incremental one, and allows the utility to improve the performance of the diagnosis system as more protective relay information becomes available. This represents an expected feature for a fault diagnosis system[15].

TABLE 1. Some of the test results of the sample power system

Test No.	The alarm signals#	Estimated faulty sections	Unobserved protective relays and their identified states
1	Operated relay: L2Rm Tripped circuit breakers: CB8, CB12	L2	L2Sp: Operated; L2Rp: Non-operated
2	Operated relay: None Tripped circuit breakers: CB7, CB11	L1	L1Sp: Operated; L1Rp: Operated
3	Operated relay: None Tripped circuit breakers: CB34, CB36	T7	T7p: Operated
4	Operated relay: None Tripped circuit breakers: CB14, CB26, CB29, CB39	T3 and L7	T3p: Operated; L7Sp: Operated; L7Rp: Operated
5	Operated relay: B1m Tripped circuit breakers: CB4, CB5, CB6, CB7, CB9, CB11	B1 and L1	L1Sp: Non-operated; L1Rp: Operated; L1Rs: Non-operated
6	Operated relays: B1m, B2m Tripped circuit breakers: CB4, CB5, CB6, CB7, CB8, CB9, CB10, CB11, CB12	B1, B2, L1 and L2	L1Sp: Non-operated; L1Rp: Operated L2Sp: Non-operated; L2Rp: Operated L1Rs: Non-operated; L2Rs: Non-operated
7	Operated relay: B1m Tripped circuit breakers: CB4, CB5, CB6, CB7, CB9, CB12, CB27	B1	L2Sp: Non-operated; L2Rp: Non-operated; L4Sp: Non-operated; L4Rp: Non-operated L2Rs: Operated; L4Rs: Operated
8	Operated relays: T7m, B7m, B8m, L5Sm, L7Rm Tripped circuit breakers: CB19, CB20, CB29, CB30, CB32, CB33, CB34, CB35, CB36, CB37 and CB39	L5, L7, B7, B8, T7 and T8	T8p: Operated; L5Rp: Non-operated L7Sp: Operated; L6Ss: Operated L8Ss: Operated
9	Operated relays: L1Sm, L7Rm, L8Sm and L8Rm Tripped circuit breakers: CB7, CB8, CB11, CB12, CB29, CB30, CB39, CB40	L1, L2, L7 and L8	L1Rp: Operated; L2Sp: Operated; L2Rp: Operated; L7Sp: Operated
10	Operated relay: B3m Tripped circuit breakers: CB7, CB12, CB14, CB15, CB19, CB31	B3	L1Sp: Non-operated; L1Rp: Non-operated L6Sp: Non-operated; L6Rp: Non-operated L1Ss: Operated; L6Rs: Operated

Only a subset of the observed protective relays, i.e., those operated protective relays, is listed in this column.

3.6 Conclusions

In this chapter, a tabu search approach for the fault section estimation and state identification of the unobserved protective relays is developed. Based upon the logic relationship among section fault, protective relay operation and circuit breaker trip, and the well developed parsimonious set covering theory, a 0-1 integer programming model describing this problem is first presented, and a tabu search method is then applied to solve the problem. The developed method can deal with any incomplete information from protective relays, and can be applied to the cases with malfunctions of protective relays and/or circuit breakers, and to any arbitrarily complicated multiple section fault cases. Extensive tests on a sample power system have demonstrated the correctness of the developed model and the efficiency of the TS based method. Moreover, the method is fast and flexible, and of potential for on-line fault diagnosis in large scale power systems.

3.7 Appendix

3.7.1 A SIMPLE EXAMPLE USED TO EXPLAIN THE DEVELOPED METHOD

The system as shown in Fig.A1 consists of 5 sections, 15 protective relays and 6 circuit breakers. The 5 sections (s_1-s_5) are A, B, C, L1 and L2, respectively. The 15 protective relays (r_1-r_{15}) are Am, Bm, Cm, L1Am, L1Bm, L2Bm, L2Cm, L1Ap, L1Bp, L2Bp, L2Cp, L1As, L1Bs, L2Bs and L2Cs. Here, m, p, s identify main protective relays, primary backup protective relays and secondary backup protective relays, respectively. Am, Bm and Cm are main protective relays of busbars A, B and C, respectively, and the other 12 protective relays are all line protective relays. For example, L2Bp represents the primary backup protective relay of line L2 at the terminal of busbar B. The operating logics of all main protective relays and all backup protective relays are listed in Table A1 and Table A2, respectively. The 6 circuit breakers (c_1-c_6) are CB1, CB2, CB3, CB4, CB5 and CB6, respectively.

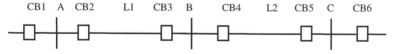

Figure A1. A Simple Example

TABLE A1.   The operating logics of main protective relays

Relay name	Operating logic
Am	If a fault occurred on A, then Am should operate to trip CB1 and CB2
Bm	If a fault occurred on B, then Bm should operate to trip CB3 and CB4
Cm	If a fault occurred on C, then Cm should operate to trip CB5 and CB6
L1Am	If a fault occurred on L1, then L1Am should operate to trip CB2
L1Bm	If a fault occurred on L1, then L1Bm should operate to trip CB3
L2Bm	If a fault occurred on L2, then L2Bm should operate to trip CB4
L2Cm	If a fault occurred on L2, then L2Cm should operate to trip CB5

TABLE A2.   The operating logics of backup protective relays

Relay name	Operating logic
L1Ap	If a fault occurred on L1 and L1Am did not operate, then L1Ap should operate to trip CB2
L1Bp	If a fault occurred on L1 and L1Bm did not operate, then L1Bp should operate to trip CB3
L2Bp	If a fault occurred on L2 and L2Bm did not operate, then L2Bp should operate to trip CB4
L2Cp	If a fault occurred on L2 and L2Cm did not operate, then L2Cp should operate to trip CB5
L1As	If a fault occurred on B, and CB3 did not trip, OR If a fault occurred on L2, and both CB3 and CB4 did not trip, THEN L1As should operate to trip CB2
L1Bs	If a fault occurred on A, and CB2 did not trip, then L1Bs should operate to trip CB3
L2Bs	If a fault occurred on C, and CB5 did not trip, then L2Bs should operate to trip CB4
L2Cs	If a fault occurred on B, and CB4 did not trip, OR If a fault occurred on L1, and both CB3 and CB4 did not trip, THEN L2Cs should operate to trip CB5

3.7.2 THE TEST SYSTEM

The sample power system as shown in Figure A2 consists of 28 sections, 84 protective relays, and 40 circuit breakers. Of the total 84 protective relays, 36 protective relays (r_1 --- r_{36}) are main relays, and the other 48 protective relays (r_{37} --- r_{84}) are backup protective relays.

S_1 --- S_{28}: A1, A2, A3, A4, T1, T2, T3, T4, T5, T6, T7, T8, B1, B2, B3, B4, B5, B6, B7, B8, L1, L2, L3, L4, L5, L6, L7, L8

C_1 --- C_{40}: CB1, CB2, CB3, ..., CB40

r_1 --- r_{36}: A1m, A2m, A3m, A4m, T1m, T2m, T3m, T4m, T5m, T6m, T7m, T8m, B1m, B2m, B3m, B4m, B5m, B6m, B7m, B8m, L1Sm, L1Rm, L2Sm, L2Rm, L3Sm, L3Rm, L4Sm, L4Rm, L5Sm, L5Rm, L6Sm, L6Rm, L7Sm, L7Rm, L8Sm, L8Rm

r_{37} --- r_{84}: T1p, T2p, T3p, T4p, T5p, T6p, T7p, T8p, T1s, T2s, T3s, T4s, T5s, T6s, T7s, T8s, L1Sp, L1Rp, L2Sp, L2Rp, L3Sp, L3Rp, L4Sp, L4Rp, L5Sp, L5Rp, L6Sp, L6Rp, L7Sp, L7Rp, L8Sp, L8Rp, L1Ss, L1Rs, L2Ss, L2Rs, L3Ss, L3Rs, L4Ss, L4Rs, L5Ss, L5Rs, L6Ss, L6Rs, L7Ss, L7Rs, L8Ss, L8Rs

Here, both A and B denote busbars; T and L denote a transformer and a transmission line, respectively; S and R denote the sending end and the receiving end of a transmission line, respectively; m, p and s denote a main protective relay, a primary backup protective relay and a secondary backup protective relay, respectively.

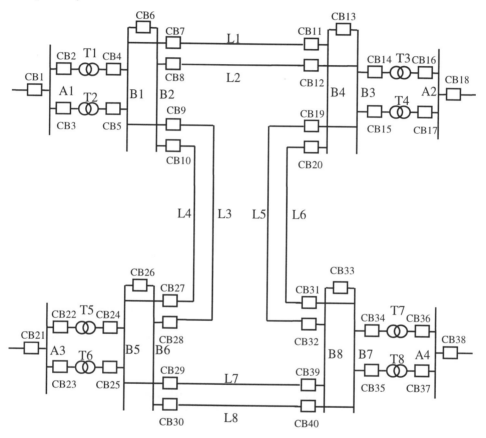

Figure A2. The sample power system

The actuating logics of each kind of protective relays are briefly introduced as follows.

a. The operation of the main protective relay of any busbar is set to trip all the circuit breakers directly connected to the busbar. For example, if a fault occurs on busbar A1, then A1m should actuate to trip CB1, CB2 and CB3; if a fault occurs on busbar B8, then, B8m should actuate to trip CB32, CB33 and CB39.

b. The operation of the main protective relays of any transformer is set to trip the two (or three, if the transformer has three windings) circuit breakers at each terminal. For example, if a fault occurs on T3, then T3m should actuate to trip CB14 and CB16.

The operation of the primary backup protective relay of any transformer is set to trip the same circuit breakers as the main protective relay only if a fault has occurred within the transformer and the circuit breakers at each terminal do not trip during the actuating time zone of its main protective relay. For example, if a fault has occurred on T3, and CB14 and/or CB16 failed to trip during the actuating time zone of T3m, then T3p should actuate to trip CB14 and/or CB16.

The purpose of the secondary backup protective relay of any transformer is to protect the transformer in case of a fault occurring on one of its neighbouring sections, and the main protective relay of the faulty section fails to make the corresponding circuit breakers trip. For example, the actuating logics of T3s are : (1) if a fault occurs on A2, and CB16 fails to trip during the actuating time zones of T3m and T3p, then T3s should actuate to trip CB16, or (2) if a fault occurs on B3, and CB14 fails to trip during the actuating time zones of T3m and T3p, then T3s should actuate to trip CB14, or (3) if a fault occurs on B4, and both CB13 and CB14 fail to trip during the actuating time zones of T3m and T3p, then T3s should actuate to trip CB14.

c. Both the sending end and the receiving end of any transmission line have one main protective relay and two backup protective relays, respectively. For example, the main protective relays of the sending end and the receiving end of L7 are L7Sm and L7Rm, respectively, and their actuating logic is: if a fault occurs on L7, then both L7Sm and L7Rm should actuate to trip CB29 and CB39, respectively.

The actuating logics of the two primary backup protective relays, i.e., L7Sp and L7Rp, are: (1) if a fault occurs on L7, and CB29 fails to trip during the actuating time zone of L7Sm, then L7Sp should actuate to trip CB29; (2) if a fault occurs on L7, and CB39 fails to trip during the actuating time zone of L7Rm, then L7Rp should actuate to trip CB39.

The purpose of the secondary backup protective relay of any transmission line is to protect the transmission line in case of a fault occurring on one of its neighbouring sections. For example, the actuating logics of L7Ss are: (1) if a fault occurs on B8, and CB39 fails to trip during the actuating time zones of L7Sm and L7Sp, then L7Ss should actuate to trip CB29, or (2) if a fault occurs on B7, and both CB33 and CB39 fail to trip during the actuating time zones of L7Sm and L7Sp, then L7Ss should actuate to trip CB29. The actuating logics of L7Rs are: (1) if a fault occurs on B5, and CB29 fails to trip during the actuating time zones of L7Rm and L7Rp, then L7Rs should actuate to trip CB39, or (2) if a fault occurs on B6, and both CB26 and CB29 fail to trip during the actuating time zones of L7Rm and L7Rp, then L7Rs should actuate to trip CB39.

3.8 Acknowledgements

One of the authors, Prof. Fushuan Wen, would like to express his sincere thanks to Zhejiang University (PRC), for permitting him to have a 1-year term of leave and to join the National University of Singapore as a Research Fellow. Partial financial support from Zhejiang Provincial Outstanding Young Scientist Fund (PRC) to Prof. Wen for this work is acknowledged.

3.9. References

1. Y.Sekine, Y.Akimoto, M.Kunugi, C.Fukui and S.Fukui, "Fault diagnosis of power systems", *Proc. of IEEE*, Vol.80, No.5, 1992, pp.673-683.
2. T.E.Dy Liacco and T.J.Kraynak, "Processing by logic programming of circuit-breaker and protective-relaying information", *IEEE PAS*, Vol.88, No.2, 1969, pp.171-175.
3. Y.M.Park, G.W.Kim, and J.M.Sohn, "A logic based expert system (LBES) for fault diagnosis of power system", *IEEE PES 96 WM 298-0 PWRS*.
4. C.Fukui and J.Kawakami, "An expert system for fault section estimation using information from protective relays and circuit breakers", *IEEE PWRD*, Vol.1, No.4, 1986, pp.83-90.
5. E. Cardozo and S.N.Talukdar, "A distributed expert system for fault diagnosis*", IEEE PWRS*, Vol.3, No.2, 1988, pp.641-646.
6. C.L.Yang, H.Okamoto, A.Yokoyama and Y.Sekine, "Expert system for fault section estimation of power system using time sequence information", *Electrical Power & Energy Systems*, Vol.14, No.2/3, 1992, pp.225-232.
7. H.Tanaka, S.Matsuda, Y.Izui and H.Taoka, "Design and evaluation of neural network for fault diagnosis", *Proc. of Second Symposium on Expert System Application to Power Systems,* Seattle, 1989, pp.378-384.
8. T.Oyama, "Fault section estimation in power system using Boltzmann machine", *Proc. of Second Forum on Artificial Neural Network Applications to Power Systems*, Japan, 1993, pp.3-8.
9. Fushuan Wen and Zhenxiang Han, "Fault section estimation in power systems using a genetic algorithm", *Electric Power Systems Research*, Vol.34, No.3, 1995, pp.165-172.
10. Fushuan Wen and Zhenxiang Han, "A refined genetic algorithm for fault section estimation in power systems using the time sequence information of circuit breakers", *Electric Machines & Power Systems,* Vol.24, No.8, 1996, pp.801-815.
11. Fushuan Wen and C.S.Chang, "A new approach to fault section estimation in power systems based on the set covering theory and a refined genetic algorithm", *Proceedings of the 12th Power Systems Computation Conference (PSCC'96),* Dresden, Germany, August 19-23, 1996, Vol.1, pp.358-365.

12. Fushuan Wen and C.S.Chang, "A new approach to fault diagnosis in electrical distribution networks using a genetic algorithm", *Artificial Intelligence in Engineering,* Vol.12, 1998, pp.69-80.
13. Fushuan Wen and C.S.Chang, "Probabilistic approach for fault-section estimation in power systems based upon a refined genetic algorithm", *IEE Proceedings: Generation, Transmission and Distribution,* Vol.144, No.2, 1997, pp.160-168.
14. Fushuan Wen and C.S.Chang, "A tabu search approach to fault section estimation in power systems", *Electric Power Systems Research,* Vol.40, No.1, 1997, pp.63-73.
15. D.S.Kirschen, B.F.Wollenberg, G.D.Irisarri, J.J.Bann and B.N.Miller, "Controlling power systems during emergencies: the role of expert systems", *IEEE Computer Applications in Power,* Vol.2, No.1, 1989, pp.41-45.
16. Y.Peng and J.A.Reggia, *Abductive Inference Models for Diagnostic Problem-solving,* Springer-Verlag, 1990.
17. F.Glover, M.Laguna, E.Taillard and D.de Werra (eds.), *Tabu search,* Science Publishers, Basel, Switzerland, 1993.
18. F.Glover and M.Laguna, *Tabu search,* Kluwer Academic Publishers, Boston, USA, 1997.
19. F.Glover, Tabu search - part I, *ORSA Journal on Computing,* Vol.1, No.3, 1989, pp.190-206.
20. F.Glover, Tabu search - part II, *ORSA Journal on Computing,* Vol.2, No.1, 1990, pp.4-32.
21. X.Bai and S.Shahidehpour, "Hydro-thermal scheduling by tabu search and decomposition method", *IEEE PWRS,* Vol.11, No.2, 1996, pp.968-974.
22. X.Bai and S.Shahidehpour, "Extended neighbourhood search algorithm for constrained unit commitment", *Electrical Power & Energy Systems,* Vol.19, No.5, 1997, pp.349-356.
23. D.Gan, Y.Hayashi and K.Nara, "Multi-level reactive resource planning by tabu search", *Proceedings of IEE of Japan Power & Energy'95, Nagoya, Japan, Aug. 2-4, 1995,* pp.137-142.
24. D.Gan, Z.Qu and H.Cai, "Large-scale var optimization and planning by tabu search", *Electric Power Systems Research,* Vol.39, No.3, 1996, pp.195-204.
25. H.T.Yang, Y.C.Huang and C.L.Huang, "Solution to capacitor placement problem in a radial distribution system using tabu search method", *Proceedings of 1995 International Conference on Energy Management and Power Delivery (EMPD'95),* Singapore, 1995, pp.388-393.
26. Y.C.Huang, H.T.Yang and C.L.Huang, "Solving the capacitor placement problem in a radial distribution system using tabu search approach", *IEEE PWRS,* Vol.11, No.4, 1996, pp.1868-1873.
27. K.Nara, Y.Hayashi, Y.Yamafuji, H.Tanaka, J.Hagihara, S.Muto, S.Takaoka and M.Sakuraoka, "A tabu search algorithm for determining distribution tie

lines", *Proceedings of 1996 International Conference on Intelligent Systems Applications to Power Systems (ISAP'96)*, Orlando, USA, 1996, pp.226-270.
28. K.Nara et al., "A new feeder route determination algorithm by Dijkstra method and tabu search", *Proceedings of 1997 International Conference on Intelligent Systems Applications to Power Systems (ISAP'97)*, Seoul, Korea, 1997, pp. 448-452.
29. H.Mori and T.Usami, "Unit commitment using tabu search with restricted neighourhood", *Proceedings of 1996 International Conference on Intelligent Systems Applications to Power Systems (ISAP'96)*, Orlando, USA, 1996, pp.422-427.
30. H.Mori et al., "An efficient method for capacitor placement with parallel tabu search", *Proceedings of 1997 International Conference on Intelligent Systems Applications to Power Systems (ISAP'97)*, Seoul, Korea, 1997, pp.387-391.
31. H.Kim, Y.Hayashi and K.Nara, "The performance of hybrized algorithm of GA, SA and TS for thermal unit maintenance scheduling, *Proceedings of 1995 IEEE International Conference on Evolutionary Computation (ICEC'95)*, Perth, Australia, 1995, Vol.1, pp.114-119.
32. Fushuan Wen and C.S.Chang, "A fuzzy abductive reasoning model for diagnostic problem solving using the tabu search approach", *Proceedings of 1997 IEEE International Symposium on Circuits and Systems (ISCAS'97)*, Vol.4, pp.2745-2748, Hong Kong, June, 1997.
33. Fushuan Wen and C.S.Chang, "Tabu search approach to alarm processing in power systems", *IEE Proceedings: Generation,Transmission and Distribution,* Vol.144, No.1, 1997, pp.31-38.
34. Fushuan Wen and C.S.Chang, "Transmission network optimal planning using the tabu search method", *Electric Power Systems Research,* Vol.42, No.2, 1997, pp.153-163.
35. R.A.Gallego, A.Monticelli and R.Romero, "Comparative studies on nonconvex optimization methods for transmission network expansion planning", *Proceedings of the 20th International Conference on Power Industry Computer Applications (PICA'97)*, May 11-16, 1997, USA, pp.24-30.

Chapter 4

GENETIC ALGORITHMS FOR SCHEDULING GENERATION AND MAINTENANCE IN POWER SYSTEMS

C.J. Aldridge, K.P. Dahal, J.R. McDonald
Centre for Electrical Power Engineering
University of Strathclyde, Glasgow, UK

4.1 Genetic Algorithms

4.1.1 INTRODUCTION

Genetic algorithms (GAs) are search and optimisation methods based on a model of evolutionary adaptation in nature. Unlike traditional 'hill-climbing' methods involving iterative changes to a single solution, GAs work with a population of solutions, which is 'evolved' in a manner analogous to natural selection. Candidate solutions to an optimisation problem are represented by chromosomes, which for example encode the solution parameters as a numeric string. The 'fitness' of each solution is calculated using an evaluation function which measures its worth with respect to the objective and constraints of the optimisation problem.

Successive 'generations' of the population are created by several simple 'genetic' operators, as illustrated in Figure 1. In each generation, solutions are selected stochastically according to their fitness in order to be recombined to form the next generation. Relatively 'fit' solutions survive, 'unfit' solutions tend to be be discarded. A new generation is created by stochastic operators - typically 'crossover', which swaps parts of binary-encoded solution strings, and 'mutation', which changes random bits in the strings. Successive generations yield fitter solutions which approach the optimal solution to the problem.

Genetic algorithms were first developed by John Holland at MIT and described in his 1975 book 'Adaptation in Natural and Artificial Systems' [1]. More recent introductory texts include those by Davis [2], Goldberg [3], Michalewicz [4] and Mitchell [5]. GAs are inherently simple, naturally parallelisable, and can generate a set of near-optimal solutions for evaluation. They provide a powerful technique to resolve complicated multi-dimensional optimisation problems, such as resource allocation and scheduling. A plethora of information and public domain GA programs are available from sites on the World Wide Web, for example [6].

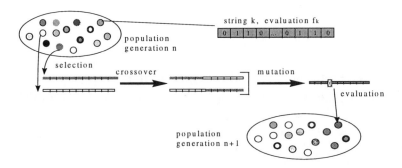

Figure 1: A generation of a basic genetic algorithm.

There are three important issues in the application of a GA to solve an optimisation problem. Firstly, how are the candidate solutions to the optimisation problem represented, in order to allow the genetic adaptation to be easily applied? Secondly, how is the optimality (quality) of the solutions assessed? Finally, how is the 'genetic adaptation' applied to existing solutions to yield new solutions? These issues are addressed in the following sections.

4.1.2 SOLUTION REPRESENTATION

GAs were initially developed using binary strings to encode the parameters of an optimisation problem. Binary encoding is a standard GA representation that can be employed for many problems: a string of bits can encode integers, real values, sets or whatever is appropriate. Furthermore, the genetic manipulation of binary chromosomes can be done by simple and universal crossover and mutation operators. However, a binary representation is often not appropriate for particular problems, and a problem-specific representation, using strings of integers or floating point numbers, character strings to represent sets, etc. may give a more coherent algorithm. Such representations require appropriately designed genetic operators. Ideally, the solution representation should be such that it represents only the feasible search space, though often this is not possible in practice.

4.1.3 EVALUATION

An evaluation function is required to assign a figure of merit (fitness) to each new solution, which should reflect the quality of the solution that the chromosome represents. During the GA 'reproduction' process the selection of individuals is done according to their fitness. If the structure of a good solution is well known it is easy to construct a suitable evaluation function. For constrained optimisation problems, the evaluation function typically comprises a weighted sum of the objective (or a simple function of it) and penalty functions to consider the constraints. This approach allows constraints to be violated, but a penalty depending on the magnitude of the violation is incurred which degrades the fitness. A highly infeasible individual has a high penalty value and will rarely be selected for reproduction, allowing the GA to concentrate on feasible or near-feasible solutions.

Multiple objectives may be included in a weighted sum in the evaluation function. However for more advanced problems, it may be useful to consider Pareto optimality [7] or fuzzy logic [8], which are outside the scope of this introduction.

4.1.4 SELECTION

In order to mimic the principle of 'survival of the fittest', GAs introduce selection pressure through choosing relatively good solutions for recombination and replacing inferior solutions in the population with new individuals. Selection is a method that stochastically picks individuals from the population according to their fitness: the higher the fitness, the more chance an individual has to be a parent. The selection pressure defines the degree to which better individuals are favoured, which drives the GA to improve the population fitness over successive generations. In general, if the selection pressure is too high, then a superior individual strongly dominates the less fit individuals and this may lead the GA to converge prematurely to a local optimum.

There are three main types of selection methods: fitness-proportionate, ranking and tournament. In fitness-proportionate selection the probability that a solution is selected is directly based on its evaluation value. In order to prevent a highly fit individual dominating the population, the evaluation values are typically scaled linearly. The 'roulette-wheel' method is the simplest and classical fitness-proportionate method. Each individual is assigned a sector of a wheel whose size is proportional to its (scaled) evaluation. A position on the wheel is chosen at random, and the individual to which that position is assigned is selected. Stochastic universal sampling (SUS) selection is similar to the roulette-wheel method, except that a set of individuals are picked simultaneously, based on a random choice of a given number of positions spaced equally around the wheel.

Ranking selection methods take account of the relative ordering of individuals with respect to their evaluation measures. The probability of selecting an individual is then given by a linear function of its rank in the population rather than its evaluation measure. This approach reduces the dominance of highly fit solutions

The basic mechanism of tournament selection involves picking a subset of individuals at random and then selecting one according to their fitness. Selection pressure is applied in choosing from the subset of individuals - for example, the best is selected with a given probability, otherwise the second best is chosen with that probability, and so on.

4.1.5 RECOMBINATION

Following their selection, 'parent' individuals are recombined to create 'offspring'. This is usually achieved using crossover and mutation operators as below, but other domain-specific operators may also be used during this process.

Crossover exploits the current solutions by exchanging elements of selected parents. This is done with a given probability, typically in the range of 0.6-1.0, otherwise parents are unchanged. One-point crossover is the simplest crossover operator, which breaks the two selected parent strings at a random position and swaps the two substrings to create two offspring which contain information from each of the parent strings. Two-point crossover is commonly used, as illustrated in Figure 2. As an alternative to such 'N-point crossover', the uniform crossover operator copies the value at each position in the off-spring from one or the other parent at random. Off-spring therefore contain a greater mixture of genetic materials from each parent.

parent strings	crossed-over strings	mutated string
101<u>010</u>110101	101<u>110</u>010101	10111001 0<u>0</u>01
001<u>110</u>011110	001<u>010</u>111110	

Figure 2: Two-point crossover and mutation of binary strings. Two crossover positions are chosen randomly (here 3 and 7) and the enclosed bits are exchanged. One of the resulting strings is randomly chosen and each bit is changed with given probability (here mutation is applied to the first string, and the tenth bit is flipped).

A mutation operator is applied to the crossed-over solutions to introduce random changes. This enables further exploration of the search space. Mutation is often seen as a background operator to maintain the genetic diversity in the population. There are many forms of mutation for different types of representation. A simple mutation operator changes the bit/value at each position in the solution string with a given small mutation probability, e.g. 0.01, as shown in Figure 2.

Mutation operators may employ hill climbing mechanisms and only apply mutation to a solution if its evaluation is improved. Such an operator can accelerate the search, but might reduce the diversity in the population and cause the algorithm converge towards some local optima.

4.1.6 POPULATION UPDATING

There are two basic population updating approaches, known as generational and steady state. The generational approach is as follows. In each generation, the population is replaced by off-spring produced by selection and recombination of parents from the population of the previous generation. The best individual in the population pool is generally retained (elitism). In this case individuals can only recombined with those from the same generation. In the alternative steady state approach, new offspring are introduced immediately into the population, replacing an existing solution, which is selected for example as the least fit or by tournament selection. Hence parents and off-springs co-exist in the population.

The recombination of two individuals is effective provided there is a sufficient diversity in the population. Ideally, the population size should be as large as possible to enhance the exploration of the search space. However, the computational time and memory required by a GA become costly as the population size increases. A population size of around 100 is typical in practice. The genetic algorithms described in the case studies below use a fixed population size, however in general this may be adapted during the course of a GA run.

An initial population of a given size must be created to begin the GA search process. The simplest way of creating the initial population is to sample the search space at random. However, heuristic methods can be used to generate some or all of the initial population. If some reasonable meaningful solutions are known or can be generated, then their inclusion in the initial population can improve the performance of the GA. An example is given in section 2 below. However the initial population should not lack diversity in order to avoid exploration of a small part of the search space.

The simplest stopping criterion is to run the GA for a fixed number of generations or iterations. Alternatively the algorithm may be continued for as long as the best solution in the population is improving or halted when the solution reaches a required quality.

Instead of a single population, a GA may use a number of smaller populations, known as 'islands'. Evolution proceeds on each island as for a single population GA, but with a regular exchange of a limited number of individuals between islands. This approach naturally lends itself to implementation on a parallel computer, with different islands allocated to individual processors [9].

4.1.7 IMPLEMENTATION

GAs are straightforward to implement for practical optimisation problems, typically requiring only the solution representation and evaluation function to be chosen. The evaluation function and genetic operators can be easily modified. GAs also yield multiple solutions which may be subsequently judged. The creation and evaluation of large number of solutions can be computationally costly, though the generational GA is naturally parallelisable.

The performance of a GA may be improved by hybridization with other solution techniques. For example, a heuristic technique may be applied to produce a meaningful initial population, as we describe below. Simulated annealing, an alternative stochastic search technique, may be combined with a GA to improve the search process [10,11]. Solutions in the final population of a GA may also be refined by an appropriate local search method.

A number of GA programs are available in the public domain, such as GENESIS [12], GENITOR [13], and RPL2 [9], which have been employed for the case studies described below. For a given application, the choice of the genetic operators and the values of parameters such as population size, crossover and mutation probabilities must generally be guided empirically.

Genetic algorithms have been applied to a range of search and optimisation problems arising in planning, scheduling and operation of power sytems. A useful bibliography is given in [14], and a recent comprehensive survey of applications of GAs and other evolutionary computing techniques in this area is given in [15]. Problems tackled include unit commitment, economic dispatch, maintenance scheduling, network expansion, alarm processing and parameter estimation. In the remainder of this chapter, we describe two case studies, in which GAs are applied to unit commitment and generator maintenance scheduling.

4.2 A Knowledge-Based Genetic Algorithm for Unit Commitment

4.2.1 INTRODUCTION

In order to meet the customer demand in a power system, the generating units must be scheduled to minimise the total cost and satisfy operating constraints. Calculating the optimal commitments (on/off) and dispatched generation for each thermal unit at a sequence of times in the scheduling period is known as the unit commitment & economic dispatch problem. This is a highly constrained combinatorial problem and continues to present a challenge for efficient solution techniques.

The constraints of the problem involve the individual units, groups of units and the entire network. Each unit is generally constrained by minimum and maximum generation, ramp rates which limit the rate of change of the generation, and minimum times that the unit can remain on or off. There may also be specified bounds on the total generation of local groups of units. The predicted demand must be met by the sum of the generation of all the units; in addition the on-line units must together maintain a specified reserve capacity. We seek the solution that satisfies these constraints and minimises the total cost, typically given by the start-up costs and running costs incurred by each unit.

The unit commitment/economic dispatch problem has been tackled using a range of solution methods. Sheble & Fahd [16] review the development of different heuristic, mathematical programming and expert system techniques over the last 30 years. Initially the commitment and dispatch problems were decoupled; indeed, the first to consider the coupled problem was Garver [17] (1963). Prior to this, and even today, priority listing is employed. In the priority list method, units are ordered according to a measure of cost and committed in this order so that their cumulative generation satisfies the required level; subsequently the dispatch of the committed units is calculated. This simple approach can however give solutions far from the optimum. The chief mathematical programming methods have been Dynamic Programming and Lagrangian Relaxation. The main drawback of Dynamic Programming is that the number of combinations of states which must be searched grows exponentially and becomes computationally prohibitive, hence methods have included for example a priority list to reduce the search space. Recent work has favoured Lagrangian Relaxation [18], in which the global constraints of demand and reserve are admitted into the objective function, and the problem decomposed into master problem and unit subproblems. This natural algorithmic decomposition admits parallelisation [19]. The method provides bounds on the original optimum but a heuristic must be employed to construct a feasible solution for the original problem.

4.2.2 GENETIC ALGORITHMS

The combinatorial aspect of the commitment problem is a natural target for the application of genetic algorithms, and in the last few years GAs have been used to solve the unit commitment/economic dispatch problem. A review of different GA approaches and results is given in [20]. Most studies have employed a single GA for the entire scheduling period. In general, the commitments are represented in the solution string as a binary array and the dispatch variables are calculated as part of the fitness function evaluation [21-30]. Representations satisfying minimum on and off times have been introduced using integers [31] and binary substrings [32,33], while in [34] both the commitment and dispatch variables were encoded in the solution string. These single GAs have included various problem-specific operators alongside the standard mutation and crossover operators. Alternatively the solution may be calculated sequentially by using a GA for each time interval in turn [35-38].

4.2.3 TEST PROBLEM FORMULATION

We consider a test problem involving 10 generating units over 24 hourly scheduling points, though our approach may easily be extended for larger problems. A minimum cost schedule is sought subject to the unit and system constraints described below. We use the following notation:

D^t	demand at time t
F_i	no-load cost for unit i
l^t	transmission constraint limit
N	number of units
R	reserve level
T	number of time intervals
U_i	start-up cost
$V_{i,1}, V_{i,2}$	incremental cost gradients
W_i	incremental cost function
x_i^t	generation of unit i at time t
x_i^{min}	minimum generation
x_i^{max}	maximum generation
x_i^*	breakpoint for piecewise linear incremental cost
α_i^t	commitment (binary)
β_i^t	start-up indictator (binary)
γ_i^t	shutdown indicator (binary)
ρ_i	ramp rate
τ_i^{on}	minimum on time
τ_i^{off}	minimum shutdown time.

Start-up and shutdown indicators are defined by

$$\beta_i^t = \begin{cases} 1 & \text{if } \alpha_i^{t-1} = 0, \ \alpha_i^t = 1, \\ 0 & \text{otherwise,} \end{cases} \qquad (1)$$

$$\gamma_i^t = \begin{cases} 1 & \text{if } \alpha_i^{t-1} = 1, \ \alpha_i^t = 0, \\ 0 & \text{otherwise.} \end{cases} \qquad (2)$$

A commitment and dispatch schedule is given by the arrays $\{\alpha_i^t\}$ and $\{x_i^t\}$, which we denote by α and x.

The total cost, composed of constant start-up costs and piecewise linear generating costs, is minimised,

$$\min_{\alpha, x} C(\alpha, x), \qquad (3)$$

where

$$C(\alpha, x) = \sum_{t=1}^{T} \sum_{i=1}^{N} \beta_i^t U_i + \alpha_i^t F_i + W_i(x_i^t), \qquad (4)$$

$$W_i = \begin{cases} x_i^t V_{i,1}, & \text{if } x_i^t \leq x_i^*, \\ x_i^* V_{i,1} + (x_i^t - x_i^*) V_{i,2}, & \text{otherwise.} \end{cases} \qquad (5)$$

subject to the following constraints:

Generation limits:
$$\alpha_i^t x_i^{\min} \leq x_i^t \leq \alpha_i^t x_i^{\max} \quad \text{for } i = 1, \ldots, N, \ t = 1, \ldots, T. \qquad (6)$$

Ramp rates:
$$\left. \begin{array}{l} \text{if } \alpha_i^{t-1} = \alpha_i^t = 1 \text{ then } -\rho_i \leq x_i^t - x_i^{t-1} \leq \rho_i, \\ \text{if } \beta_i^t = 1 \text{ then } x_i^t = x_i^{\min}, \\ \text{if } \gamma_i^t = 1 \text{ then } x_i^{t-1} \leq x_i^{\min} + \rho_i, \end{array} \right\} \quad \text{for } i \in I_{RR}, \ t = 1, \ldots, T. \qquad (7)$$

Minimum on times:
$$\beta_i^t + \sum_{t'=t+1}^{\min(t+\tau_i^{on}-1, T)} \gamma_i^t \leq 1 \quad \text{for } i \in I_{MOT}, \ t = 1, \ldots, T-1. \qquad (8)$$

Minimum shutdown times:
$$\gamma_i^t + \sum_{t'=t+1}^{\min(t+\tau_i^{off}-1, T)} \beta_i^t \leq 1 \quad \text{for } i \in I_{MST}, \ t = 1, \ldots, T-1. \qquad (9)$$

Demand:
$$\sum_{i=1}^{N} x_i^t = D^t \quad \text{for } t = 1, \ldots, T. \qquad (10)$$

Reserve:
$$\sum_{i=1}^{N} \alpha_i^t x_i^{\max} \geq D^t + R \quad \text{for } t = 1, \ldots, T. \qquad (11)$$

Transmission constraint:

$$\sum_{i \in I_{TC}} x_i^t \geq l^t \quad \text{for } t \in T_{TC}. \tag{12}$$

Initial conditions:
$$\alpha_i^0, x_i^0 \text{ given for } i = 1, \ldots, N. \tag{13}$$

In the above I_{RR}, I_{MOT}, I_{MST} and I_{TC} denote particular subsets of units associated with the constraints, and T_{TC} is a subset of times. Equations (1)-(13) define a mixed integer programming problem. This may be made linear by formulating (1),(2) and (7) as inequalities and introducing extra variables to reformulate (5). In this form the problem is amenable to Lagrangian relaxation.

The parameter values for the units and demand profile are given in Tables 1 and 2. In addition we take
$$R = 200 \text{ MW}, \; l^t = 1600 \text{ MW}, \; I_{TC} = \{1,4,7\}, \; T_{TC} = \{20,21,22\},$$
and initial conditions
$$x_1^0 = 300 \text{ MW}, \; x_2^0 = 700 \text{ MW}, \; x_3^0 = 900 \text{ MW}, \; \alpha_4^0 = \ldots = \alpha_{10}^0 = 0.$$

In order to gauge the computation required to calculate the optimal solution to this problem, a series of similar problems with fewer constraints and time intervals were solved using branch-and-bound. These problems are given by objective function (3)-(5), initial conditions (13) and generation limits (6), plus (a) demand constraint (10), (b) demand and reserve constraints (10) and (11), and (c) all constraints (7)-(12), Branch-and-bound was applied to these problems over the first T time intervals, for T=2,4,8,12,24, using standard OSL [39] subroutines on a Sun Sparc workstation. The CPU time to calculate the optimal solution to problem (a) with T=24 was over 10 hours; problem (b) with T=12 required over 6 hours. The solution to problem (b) with T=24 and problem (c) with T=8 was not found within 12 hours of CPU time.

4.2.4 GENETIC ALGORITHM DESIGN

4.2.4.1 *Solution Representation and Evaluation*
We consider the unit commitment/economic dispatch problem (3)-(13) in the decomposed form:
$$\min_{\alpha} F(\alpha), \tag{14}$$
subject to (8),(9), (11), and the following problem being feasible:

$$F(\alpha) = \min_{x} \{C(\alpha, x): x \text{ satisfies } (6), (7), (10) \text{ and } (12)\}. \tag{15}$$

Table 1: Generating unit data, where '-' indicates that the corresponding constraint is not specified for that unit, and units other than 2 and 5 have $x_i^* = x_i^{max}$.

i	x_i^{min} (MW)	x_i^{max} (MW)	ρ_i (MW/h)	τ_i^{on} (h)	τ_i^{on} (h)	U_i (£)	F_i (£/h)	$V_{i,1}, V_{i,2}; x_i^*$ (£/MWh);
1	300	1000	40	6	2	14,000	5000	10
2	300	1000	180	6	-	14,000	7875	3.75,15; 700
3	400	1000	600	6	-	20,000	9500	5
4	150	500	60	2	2	10,000	3750	15
5	150	500	240	2	-	10,000	5906.25	5.625,22.5;
6	200	500	300	2	-	25,000	7125	7.5
7	200	200	-	-	2	2000	2000	30
8	200	200	-	-	-	1200	1200	31
9	100	200	-	-	-	800	800	35
10	100	200	-	-	-	0	0	40

Table 2: Demand profile.

t	D^t (MW)	0700	3000	1400	3500	2100	3500
0100	2400	0800	4100	1500	3200	2200	2700
0200	2200	0900	4150	1600	3700	2300	2200
0300	2000	1000	4200	1700	4500	2400	1900
0400	1850	1100	4250	1800	5050		
0500	1750	1200	4300	1900	4700		
0600	1700	1300	4000	2000	4200		

A GA is applied to the combinatorial minimisation unit commitment problem (14). The continuous economic dispatch problem (15) parameterised by α is solved in the evaluation function of the GA, and this may be done by linear programming.

The solution representation in the GA is therefore the binary commitment matrix α. In the implementation of the GA this was stored as a binary string consisting of the commitments ordered by time periods first and units second,

$$(\alpha_1^1, \ldots, \alpha_N^1, \ldots, \alpha_1^T, \ldots, \alpha_N^T). \quad (16)$$

Each string is evaluated by first solving (15) to give x, and then summing the total cost given by (4) and penalty functions for violations of constraints (6)-(12). From (6) and (10) a necessary condition on α is

$$\sum_{i=1}^{N} \alpha_i^t x_i^{min} \geq D^t \text{ for } t = 1, \ldots, T. \quad (17)$$

The evaluation function is taken as

$$f(\alpha) = C(\alpha, x) + w_a P_a(\alpha) + w_b P_b(\alpha) + w_c P_c(\alpha) + w_d P_d(\alpha) + w_e P_e(\alpha, x).$$
(18)

Here P_a is a penalty function associated with constraints (8) and (9), P_b with (11), P_c with (17), P_d with (10) and (12), P_e with (6) and (7), and the w_a etc. are weights. The penalty functions increase linearly with the constraint violations, and are chosen with the weights so that the penalty terms are typically larger than the cost terms.

4.2.4.2 Population Updating

An initial population of K solutions $\{\alpha^{(1)}, \ldots, \alpha^{(K)}\}$ is created. These are chosen randomly, or else the initial population is 'seeded' using the method described below. The evaluation value $f^{(k)}$ of each solution is then calculated. This may be done by linear programming, but as explained below a heuristic method is used to approximate $f^{(k)}$. A new population is created in the following steps.

1. The lowest evaluation solution $\alpha^{(k*)}$ is copied to the new population (elitism).
2. A set of $2(m-1)$ parent solutions are selected from the old population by stochastic universal sampling [4]. This is done by ranking the solutions in order of increasing $f^{(K)}$, so $\alpha^{(k*)}$ has rank 1. Solutions are then selected in proportion to a decreasing linear function of their rank.
3. A pair of parent solutions are combined by one-point crossover with probability p_c to create a new solution. Bit-wise mutation is then applied to the new solution with mutation rate p_m. The mutated solution is then evaluated and placed in the new population. This crossover, mutation and evaluation is done K-1 times to complete the population in the next generation of the GA.

The population updating is repeated for J generations, using the heuristic method for solution evaluation. The best solution in the final population is then re-evaluated by calculating $f^{(k^*)}$ exactly by linear programming.

4.2.4.3 Selection of Initial Population

A method to identify the likely structure of the unit commitments was derived following knowledge elicitation with scheduling experts, and the construction and validation of a knowledge model. This was done using the KADS (Knowledge Acquisition and Design Structuring) methodology [40].

Typically a number of units are committed throughout the scheduling period, while others remain uncommitted. These groups of units may be heuristically determined, largely by operating cost; however, the inflexibility of certain units and the transmission constraints must also be taken into account.

The units are initially placed in merit order (here in order of increasing running cost/MW at full output), and their cumulative total generations calculated. Units which lie sufficiently (say $\geq m$ MW) below the minimum demand are classified as 'must-run'; units which lie sufficiently above the maximum demand are classified as 'can't-run'; and those remaining are 'can-run' units. This classification is then revised at each time interval to take account of unit inflexibilities and transmission constraints - in this case, constraint (12) - and the can-run band subsequently narrowed to a margin of around the demand curve. Here we use a margin of width m=500 MW.

The resulting 'partition' may then be used to initialise the population of the GA. For each solution in the initial population, the commitments are then set as $\alpha_i^t = 1$ (must-run), $\alpha_i^t = 0$ (can't-run), or chosen randomly (can-run).

4.2.4.4 Heuristic Evaluation

The evaluation of each commitment string requires the solution of the economic dispatch problem (15). In order to realise an efficient algorithm, a fast heuristic method was used to solve (15) rather than a standard linear programming solver. An existing rule-based method for generation scheduling (commitment and dispatch) was identified and described in a knowledge model, again using the KADS methodology. From this model a heuristic method was derived for economic dispatch with commitments given.

In this method the x_i^t are calculated at a sequence of time intervals, ordered according to the maxima and minima of the demand profile. At each time interval the x_i^t are successively decreased and increased, using merit order, in order to satisfy the group constraints and demand. This is done taking into account the values of x_i^t at previously set times, the given commitments, and the unit capacities and ramp-rates. This approximate method proved to a fast and sufficiently accurate alternative to an exact linear programming method.

To gauge the effectiveness of these knowledge-based methods, the GA was initially applied to a smaller, simple problem with a known optimum solution. Results showed that choosing the initial population based on the derived partition and using the heuristic dispatch method in the evaluation function significantly reduced the computational time of the GA to find the optimum, compared to a GA with random initial population and exact LP evaluation [21]. A schematic of the augmented GA is shown in Figure 2.

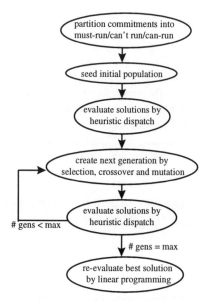

Figure 2: Schematic of knowledge-based GA.

4.2.5 RESULTS

The GA was implemented using RPL2 [9] and the LP re-evaluation was done using AMPL [41] with OSL solver routines on a Sun Sparc5 workstation. The results presented in this section were obtained with the GA parameters $J=1000$, $K=100$, $p_c = 0.9$, $p_m = 0.0015$. These values were found to give the best results in a study of the sensitivity of the GA performance on the above parameters.

The GA was applied to the test problem and the results compared with those obtained using Lagrangian relaxation (LR). Table 3 shows the cost of the solution obtained using LR with 10 sub-gradient iterations, the average cost over ten independent runs of the GA solution, and the associated CPU times. A lower bound, given by the dual cost after 200 iterations, is also shown, which is useful in assessing solutions. The GA cost is 5.7% higher than the lower bound, compared to 6.6% for LR.

4.2.6 CONCLUSIONS

A knowledge-based genetic algorithm has been developed for the unit commitment/ economic dispatch problem. In the GA each binary string is a complete commitment schedule, and the corresponding dispatched generations are calculated in the evaluation of each string. Expert knowledge of generation scheduling has been modelled and used to define the initial conditions of the GA. This has been shown to significantly improve the convergence. Scheduling rules have been incorporated in a fast approximate method of evaluating solutions, accelerating the computational time of the GA to competitive levels. The knowledge-based genetic algorithm has been applied to a representative test problem and shown to obtain better solutions than Lagrangian relaxation (LR) in similar computational times.

Table 3: Comparison of results given by knowledge-based genetic algorithm and Lagrangian relaxation.

LR	solution cost	$1.5603 * 10^6$
	CPU time	145 s
	lower bound	$1.4638 * 10^6$
GA	solution cost	$1.5470 * 10^6$
	CPU time	257 s
	improvement on LR	0.38%

4.3 Generator Maintenance Scheduling Using a Genetic Algorithm

4.3.1 INTRODUCTION

It is vital for a utility to determine when its generators should be taken off-line for preventive maintenance. This is primarily because other short-term and long-term planning activities such as unit commitment, generation dispatch, import/export of power and generation expansion planning are directly affected by such decisions. In modern power systems the demand for electricity has greatly increased with related expansions in system size, which has resulted in higher numbers of generators and lower reserve margins making the generator maintenance scheduling (GMS) problem more complicated. The goal of GMS is to calculate a maintenance timetable for generators in order for example to maintain a high system reliability, reduce total operation cost, and extend generator life time, while satisfying constraints on the individual generators and the power system.

There are generally two categories of objectives in GMS, based on reliability [8,42-48] and economic cost [10,11,45,47-50]. The levelling of the reserve generation over the entire operational planning period is the most common reliability criterion. This can be realised by maximising the minimum net reserve of the system during any time period [45,46,48]. In the case of a large variation of reserve, minimising the sum of squares of the reserves can be an effective approach [46]. Alternatively, the quality of reserve is considered, whereby the risk of exceeding the available capacity is levelled over the entire period by using the equivalent load carrying capacity for each unit and an equivalent load for each interval [46,48]. Minimising the sum of the individual loss of load probabilities for each interval can also be a reliability objective under the conditions of load uncertainty and random forced outages of units [46].

The most common economic objective is to minimise the total operating cost, which includes the costs of energy production and maintenance. If outage durations are allowed to vary, this results in a trade-off solution between the energy production cost and the maintenance cost. Shorter outage durations lead to higher maintenance costs but reduce the load of expensive generation and possible energy purchases, resulting in lower energy production costs [47]. The production cost alone could also be chosen as the objective function by minimising the total energy replacement cost due to preventive maintenance scheduling. However, this is an insensitive objective as it requires many approximations [47,48].

Any maintenance timetable must satisfy a given set of constraints. Typical constraints of the GMS problem are:

- Maintenance window constraints, which define the possible times and the duration of maintenance for each unit.
- Crew constraints, which consider the manpower availability for maintenance work.
- Resource constraints, which specify the limits on the resources needed for maintenance at each period.
- Exclusion constraints, which prevent the simultaneous maintenance of a set of units.
- Sequence constraints, which restrict the initiation of maintenance of some units after a period of maintenance of some other units.
- Load constraints, which consider the demand on the power system during the scheduling period.
- Reliability constraints, which consider the risk level of a given maintenance schedule.
- Transmission capacity constraints, which specify the limit of transmission capacity in an interconnected power system.
- Geographical constraints, which limit the number of generators under maintenance in a region.

In general GMS is a multi-criterion constrained combinatorial optimisation problem, with nonlinear objective and constraint functions. Several deterministic mathematical methods and simple heuristic techniques are reported in the literature for solving particular GMS problems [45,46,47,48]. Mathematical methods are based on integer programming, branch-and-bound and dynamic programming. However these methods are unsuitable for the nonlinear objectives and constraints of GMS and their computational time grows prohibitively with problem size. The heuristic methods use a trial-and-error method to evaluate the maintenance objective function in the time interval under examination. They require significant operator input and may even fail to find feasible solutions [46,47].

In order to overcome the above limitations a number of artificial intelligence approaches for GMS have been studied [43]. Genetic algorithms (GAs) offer an effective alternative method to solve complex combinatorial optimisation problems, and have recently been applied to GMS using binary strings to represent the maintenance timetable [10,11,42,50] and integer representation [8,42,43]. In all cases, penalty functions were used in the formulation of the evaluation function to take account of violations of problem constraints. GAs have been hybridized with other techniques in order to include scheduling heuristics and improve the performance of the solution algorithm. In [8,49] fuzzy logic was used in the evaluation of each candidate solution, in order to model flexibilities in scheduling using expert knowledge. A knowledge-based technique was employed in [49] for load flow calculation within the evaluation function to improve the speed of the algorithm. GAs have been applied to GMS using the acceptance probability of the simulated annealing (SA) method for the survival of a candidate solution during the evolution process [10,11]. If a newly created solution is an improvement, it is accepted, otherwise it is accepted with a defined probability. The hybridization improved the convergence of the algorithms. In [10] a tabu search (TS) technique was also coupled with the GA/SA hybrid method. In each generation, the best solution was selected as the new trial solution for the TS to improve the search in the neighbourhood of the solution.

In the following sections we describe an application of GAs to a GMS test problem. Both steady state and generational GAs are employed using an integer representation. The GMS test problem and its mathematical model are described in section 4.2. Section 4.3 details the implementation of the genetic algorithm technique to the problem. The test results and the performances of the GAs are discussed in section 4.4, and our conclusions follow in section 4.5.

4.3.2 TEST GMS PROBLEM

A test problem of scheduling the maintenance of 21 units over a planning period of 52 weeks is considered, which is loosely derived from the example presented in [47] with some simplifications and additional constraints. The problem involves the reliability criterion of minimising the sum of squares of the reserves in each week. Each unit must be maintained (without interruption) for a given duration within a specified window, and the available manpower is limited. Table 4 gives the capacities, allowed periods and duration of maintenance and the manpower required for each unit. The system's peak load is 4739 MW, and there are 20 people available for maintenance work in each week.

Table 4: Data for the test system.

Unit	Capacity (MW)	Allowed period	Outage (weeks)	Manpower required for each week
1	555	1-26	7	10+10+5+5+5+5+3
2	555	27-52	5	10+10+10+5+5
3	180	1-26	2	15+15
4	180	1-26	1	20
5	640	27-52	5	10+10+10+10+10
6	640	1-26	3	15+15+15
7	640	1-26	3	15+15+15
8	555	27-52	6	10+10+10+5+5+5
9	276	1-26	10	3+2+2+2+2+2+2+2+2+3
10	140	1-26	4	10+10+5+5
11	90	1-26	1	20
12	76	27-52	3	10+15+15
13	76	1-26	2	15+15
14	94	1-26	4	10+10+10+10
15	39	1-26	2	15+15
16	188	1-26	2	15+15
17	58	27-52	1	20
18	48	27-52	2	15+15
19	137	27-52	1	15
20	469	27-52	4	10+10+10+10
21	52	1-26	3	10+10+10

The GMS problem can be formulated as an integer programming problem by using binary variables, either indicating the period in which maintenance of each unit starts [8,42-45,50] or representing the maintenance status of each unit at each time [10,11,46-48]. The variables in the first formulation are bounded by the maintenance window constraints and hence the search space is reduced. The test problem is formulated below using these variables. We introduce the following notation:

i	index of generating units
I	set of generating unit indices
N	total number of generating units
t	index of periods
T	set of indices of periods in planning horizon
e_i	earliest period for maintenance of unit i to begin
l_i	latest period for maintenance of unit i to end
d_i	duration of maintenance for unit i

P_{it} generating capacity of unit i in period t
L_t anticipated load demand for period t
M_{it} manpower needed by unit i at period t
AM_t available manpower at period t

Suppose $T_i \subset T$ is the set of periods when maintenance of unit i may start, so $T_i = \{t \in T: e_i \leq t \leq l_i - d_i + 1\}$ for each i. We define

$$X_{it} = \begin{cases} 1 & \text{if unit i starts maintenance in period t,} \\ 0 & \text{otherwise,} \end{cases}$$

to be the maintenance start indicator for unit $i \in I$ in period $t \in T_i$. It is convenient to introduce two further sets. Firstly let S_{it} be the set of start time periods k such that if the maintenance of unit i starts at period k that unit will be in maintenance at period t, so $S_{it} = \{k \in T_i: t - d_i + 1 \leq k \leq t\}$. Secondly, let I_t be the set of units which are allowed to be in maintenance in period t, so $I_t = \{i: t \in T_i\}$. Then the problem can be formulated as a quadratic 0-1 programming problem as below.

The objective is to minimise the sum of squares of the reserve generation

$$\underset{X_{it}}{\text{Min}} \left\{ \sum_{t \in T} \left(\sum_{i \in I} P_{it} - \sum_{i \in I_t} \sum_{k \in S_{it}} X_{ik} P_{ik} - L_t \right)^2 \right\}, \quad (19)$$

subject to the maintenance window constraint

$$\sum_{t \in T_i} X_{it} = 1 \quad \text{for all } i \in I, \quad (20)$$

the manpower constraint

$$\sum_{i \in I_t} \sum_{k \in S_{it}} X_{ik} M_{ik} \leq AM_t \quad \text{for all } t \in T, \quad (21)$$

the load constraint

$$\sum_i P_{it} - \sum_{i \in I_t} \sum_{k \in S_{it}} X_{ik} P_{ik} \geq L_t \quad \text{for all } t \in T. \quad (22)$$

4.3.3 GA IMPLEMENTATION

A solution to the test problem may be represented as a one-dimensional binary string which consists of sub-strings X_{i,e_i}, X_{i,e_i+1}, ..., X_{i,l_i-d_i+1} for each unit i. The size of the GA search space for this type of representation is

$$\sum_{2\,i=1}^{N}(l_i-d_i-e_i+2)$$

For each unit i=1,2,...,N, the maintenance window constraint (20) forces exactly one variable in $\{X_{it}: t \in T_i\}$ to be one and the rest to be zero. The solution of the problem thus amounts to finding the correct choice of positive variable from each variable set $\{X_{it}: t \in T_i\}$, for i=1,2,...,N. The index t_i^* of this positive variable indicates the period when maintenance for unit i starts. In order to reduce the number of variables the t_i^*, i=1,2,...,N, can be taken as new variables. These can be expressed as binary numbers, in a 'binary for integer' representation. However, a direct integer representation automatically considers the maintenance window constraint (20) and greatly reduces the size of the GA search space to

$$\prod_{i=1}^{N}(l_i - d_i - e_i + 2).$$

We present results obtained using the integer representation, which has been found to give significantly better results than the binary representation or binary for integer representation [44].

The merit of the solution represented by the GA string is calculated by an evaluation function, given by a weighted sum of the objective and penalty functions for violations of the constraints, which we seek to minimise. The penalty value for each constraint violation is proportional to the amount by which the constraint is violated, hence

$$\text{evaluation} = \omega_O \text{ SSR} + \omega_M \text{ TMV} + \omega_L \text{ TLV}, \qquad (23)$$

where SSR is the sum of squares of reserves as in (19), TMV is the total manpower violation of (21), and TLV is the total load violation of (22). The weighting coefficients ω_O, ω_M and ω_L are chosen so that the penalty values for the constraint violations dominate over the objective function, and the violation of the relatively hard load constraint (22) gives a greater penalty value than for the relatively soft crew constraint. This is because a solution with a high reliability but requiring more manpower may well be accepted for a power utility as the unavailable manpower may be hired. In fact, there is a trade-off between the level of reliability and the required extra manpower. This flexibility of the problem can be modelled using a fuzzy logic approach within the evaluation function [8].

4.3.4 TEST RESULTS AND DISCUSSION

Both generational (GN) and steady state (SS) GAs were implemented for the test problem using tournament selection, two-point crossover, random mutation and elitism. A tournament replacement operator was employed for the SS GA. GAs were implemented for the test problem using the RPL2 program [9] on a Sun Sparcstation 1000.

The performance of a GA is generally dependent on the GA parameters used, in particular the crossover and mutation probabilities. The sensitivity of both GAs to the variation of crossover and mutation probabilities CP and MP in the range of 0.6-1.0 and 0.001-0.1 respectively was therefore established. The results are depicted in Figure 3, which shows for each case the average evaluation value of the best solutions obtained from ten independent GA runs. A different initial population was randomly created for each run but the same ten initial populations were used for each case. The total number of iterations (solutions created) in each run was fixed as 30,000. The population size was taken to be 100.

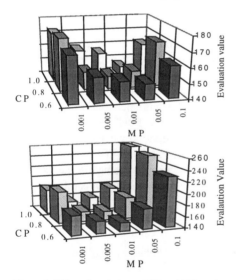

Figure 3: Effect of variations of CP and MP on the performance of SS GA (left) and GN GA (right).

Table 5: Best results obtained from SS GA and GN GA.

	SS GA	GN GA
CP, MP	1.0, 0.05	0.6, 0.01
average evaluation value (over ten runs)	146.71	155.05
best evaluation value (over ten runs)	137.91	148.31
CPU time (one run)	34s	25s

As Figure 3 shows, the average results of both GAs do not vary greatly for varying CP, but are more sensitive to variations of MP, particularly for the GN GA. The SS GA gives the best performance at higher crossover and mutation probabilities than the GN GA. The best results of both GAs are given in Table 5. Hence the SS GA finds better schedules (with lower evaluation values) than the GN GA. However the CPU time (which increases as MP increases) for the GN GA is smaller than that for the SS GA. For both GAs, the best solution (over ten runs) is feasible, so the values shown in Table 5 represent the objective value (SSR multiplied by weighting coefficient ω_O).

The best solution found by the SS GA, whose evaluation measure is 137.91, is illustrated in Figure 4 (left). The schedule represented by the solution is set out in the top portion of the figure, in which the horizontal bars indicate the maintenance of a generating unit. The middle portion of the figure shows the reserve margins in each week for the schedule, which are non-negative since the schedule satisfies the load constraint. The manpower requirements in each week for the solution are depicted in the bottom portion of the figure, which are within the available level.

In order to compare with the best GA solution, we developed a solution heuristically by timetabling the maintenance outages of generators in order of decreasing capacity, to level the reserve generation while considering the maintenance window and load constraints. The schedule, reserve margins and manpower requirements for each week given by the heuristic solution is illustrated in Figure 4 (right). The solution respects the load constraints but violates the manpower constraints in three time periods. The evaluation value of the solution is 222.61, which is the weighted sum of the objective value (134.61) and the amount of the violation of the constraints. Hence the objective value of the heuristic solution is better than that of the best GA solution, but the solution is infeasible.

The convergence of the SS GA in finding the best solution (with CP=1.0, MP=0.05) is depicted in Figure 5, which shows the evaluation value of the best solution found so far and the mean evaluation value of the solutions in the population against the number of iterations.

The reduction of the mean evaluation value and the evaluation value of the best solution is very quick in the initial stage, up to 5000 iterations, of the GA. During this stage, the GA mainly concentrates on finding feasible solutions to the problem. The population does not converge to the best solution even after a large number of iterations as the high mutation probability (0.05) maintains the diversity in population. The convergence of the algorithm can be improved even with a high mutation probability if the probabilistic acceptance criteria of a SA technique is incorporated into the GA method. A further improvement can be gained by initialising the population using a heuristic schedule. This results will be reported elsewhere.

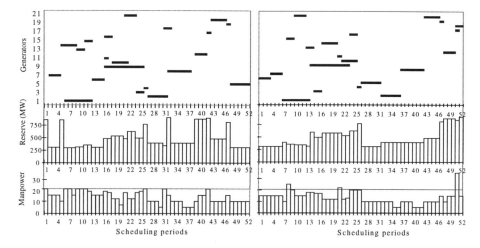

Figure 4: The schedule given by the best GA solution (left) and the heuristically developed schedule (right).

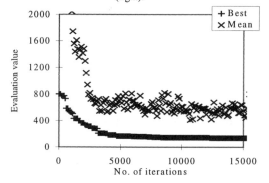

Figure 5: Performance of the SS GA in finding the best solution.

4.3.5 CONCLUSIONS

A GA technique using an integer representation has been demonstrated for a test problem of generator maintenance scheduling. The use of an integer rather than binary representation greatly reduces the GA search space and is straightforward to implement. A penalty function approach has been employed to consider the constraints of the problem. Two GAs with steady state and generational design were tested and the effect of varying crossover and mutation probabilities were studied. The test results show that both the GAs are stable to variation in crossover probability in the expected range. The GN GA is found more sensitive to variation in mutation probability than the SS GA. The integer SS GA gives better performance than the integer GN GA in term of finding better solution in a fixed number of iterations, but the latter is found to be faster.

The results presented above show that the GA is a robust and stable technique for the solution of GMS problems. Good solutions of the problem can be found if an appropriate problem encoding, GA approach, evaluation function and GA parameters are selected for the problem.

4.4 Acknowledgements

This work was carried out in the Rolls-Royce University Technology Centre in Power Engineering at the University of Strathclyde. The first author was supported by the Engineering and Physical Science Research Council and The National Grid Company, and the second author was supported by Rolls-Royce plc. The authors also acknowledge the use of the Reproductive Plan Language, RPL2, produced by Quadstone Limited, in the production of this work.

4.5 References

1. Holland, J.H (1975) *Adaptation in Natural and Artificial Systems*, University of Michigan Press.
2. Davis, L. (1991) *Handbook of Genetic Algorithms*, Van Nostrand Reinhold.
3. Goldberg, D.E. (1989) *Genetic Algorithms in Search, Optimisation, and Machine Learning*, Addison-Wesley.
4. Michalewicz, Z. (1994) *Genetic Algorithms + Data Structures =Evolution Programs*, Springer-Verlag.
5. Mitchell, M. (1996) *An Introduction to Genetic Algorithms*. MIT Press.
6. *GA Archive*, web site http://www.aic.nrl.navy.mil/galist/
7. Fonseca, C.M. and Fleming, P.J. (1993) Genetic algorithms for multiobjective optimization: formulation, discussion and generalization, *Proceeding of the 5th International Conference on Genetic Algorithms (ICGA'93)*, Morgan Kaufmann Publishers, 416-423.
8. Dahal, K.P., Aldridge, C.J. and McDonald, J.R. (in press) Generator maintenance scheduling using a genetic algorithm with a fuzzy evaluation function, *Fuzzy Sets and Systems*.
9. Quadstone Limited (1997) Reproductive Plan Language (RPL2), User manual.
10. Kim, H., Hayashi, Y. and Nara, K. (1997) An algorithm for thermal unit maintenance scheduling through combined use of GA, SA and TS, *IEEE Transactions on Power Systems* **12,** 329-335.
11. Kim, H., Nara, K. and Gen, M. (1994) A method for maintenance scheduling using GA combined with SA, *Computers and Industrial Engineering* **27,** 477-480.

12. Grefenstette, J. (1990) *A user's guide to GENESIS*. ftp.aic.nrl.navy.mil/pub/galist/src/ga.
13. Whitley, D.L. (1990) *GENITOR*. Ftp.cs.colostate.edu/pub/GENITOR.tar.
14. Alander, J.T. (1996) *An indexed bibliography of genetic algorithms in power engineering*. Report 94-1-POWER, University of Vaasa, ftp.uwasa.fi,/cs/report94-1/gaPOWERbib.ps.Z.
15. Miranda, V., Srinivasan, D. and Proença, L.M. (1998) Evolutionary computation in power systems, *Electrical Power & Energy Systems* **19**, 45-55.
16. Sheble, G.B. and Fahd, G.N. (1994) Unit commitment literature synopsis, *IEEE Transactions on Power Systems* **9**, 128-135.
17. Garver, L. (1963) Power generation scheduling by integer programming --- development of theory, *AIEE Transactions* **81**, 1212-1218.
18. Oliveira, P., McKee, S., and Coles, C. (1992) Lagrangian relaxation and its application to the unit-commitment-economic-dispatch problem, *IMA Journal of Mathematics Applied in Business and Industry* **4**, 261-272.
19. Oliveira, P., Blair-Fish, J., McKee, S., and Coles, C. (1992) Parallel Lagrangian relaxation in power scheduling, *Computer Systems in Engineering* **3**, 609-612.
20. Aldridge, C.J., McKee, S. and McDonald, J.R. (1997) Genetic algorithm methodologies for scheduling electricity distribution, in M. Brøns, M.P. Bendsøe and M.P. Sørensen (eds.), *Progress in Industrial Mathematics at ECMI'96*, Teubner, 364-371.
21. Aldridge, C.J., McDonald, J.R. and McKee, S. (1997) Unit commitment for power systems using a heuristically augmented genetic algorithm, *Proceedings of 2nd International Conference on Genetic Algorithms in Engineering Systems: Innovations and Applications (GALESIA'97)*, IEE Conference Publication 446, 433-438.
22. Cai, X.-Q. and Lo, K.-M. (1997) Unit commitment by a genetic algorithm, *Nonlinear Analysis, Theory, Methods & Applications* **30**, 4289-4299.
23. Hassoun, M.H. and Watta, P. (1994) Optimization of the unit commitment problem by a coupled gradient network and by a genetic algorithm, *report no. TR-103697*, Electric Power Research Institute.
24. Kazarlis, S.A., Bakirtzis, A.G. and Petridis, V. (1996) A genetic algorithm solution to the unit commitment problem, *IEEE Transactions on Power Systems* **11**, 83-90.
25. Ma, X., El-Keib, A.A., Smith, R.E. and Ma, H. (1995) A genetic algorithm based approach to thermal unit commitment of electrical power systems, *Electrical Power Systems Research* **34**, 29-36.
26. Maifeld, T.T. and Sheble, G.B. (1996) Genetic-based unit commitment algorithm, *IEEE Transactions on Power Systems* **11**, 1359-1370.
27. Numnonda, T., Annakkage, U.D. and Pahalawaththa, N.C. (1996) Unit commitment using stochastic optimisation, *Proceedings of Intelligent Systems Applications in Power Systems (ISAP'96)*, 429-433.

28. Orero, S.O. and Irving, M.R. (1997) A combination of the genetic algorithm and Lagrangian relaxation decomposition techniques for the generation unit commitment problem, *Electrical Power Systems Research* **43**, 149-156.
29. Sheble, G.B. and Maifeld, T.T. (1994) Unit commitment by geneticalgorithm and expert system, *Electrical Power Systems Research* **30**, 115-121.
30. Shebleé, G.B., Maifeld, T.T., Brittig, K., Fahd, G. and Fukurozaki-Coppinger, S. (1996) Unit commitment by genetic algorithm with penalty methods and a comparison of Lagrangian search and genetic algorithm - economic dispatch example, *Electrical Power & Energy Systems* **18**, 339-346.
31. Saitoh, H., Inoue, K. and Toyoda, J. (1994) Genetic algorithm approach to unit commitment, in A. Hertz, A.T. Holen and J.C. Rault (eds.) *Proceedings of the International Conference on Intelligent System Application to Power Systems*, 583-589.
32. Yang, P-C., Yang, H-T. and Huang, C-L. (1996) Solving the unit commitment problem with a genetic algorithm through a constraint satisfaction technique, *Electrical Power Systems Research* **37**, 55-65.
33. Yang, H.-T., Yang, P.-C. and Huang, C.-L. (1997) A parallel genetic algorithm approach to solving the unit commitment problem: implementation on the transputer networks, *IEEE Transactions on Power Systems* **12**, 661-668.
34. Oliveira, P., McKee, S., and Coles, C. (1994) Genetic algorithms and optimising large nonlinear systems, in J.H. Johnson, S. McKee and A. Vella (eds.) *Artificial Intelligence in Mathematics*, OUP, 305-312.
35. Dasguptar, D. and McGregor, D.R. (1994) Thermal unit commitment using genetic algorithms, *IEE Proceedings C - Generation, Transmission and Distribution* **141**, 459-465.
36. Orero, S.O. and Irving, M.R. (1995) Scheduling of generators with a hybrid genetic algorithm, *Proceedings of 1st International Conference on Genetic Algorithms in Engineering Systems: Innovations and Applications (GALESIA'95)*, IEE Conference Publication no. 414, 200-206.
37. Orero, S.O. and Irving, M.R. (1996) A genetic algorithm for generator scheduling in power system, *International Journal of Electrical Power and Energy Systems* **18**, 19-26.
38. Orero, S.O. and Irving, M.R. (1997) Large scale unit commitment using a hybrid genetic algorithm, *International Journal of Electrical Power and Energy Systems* **19**, 45-55.
39. IBM (1992) *Optimisation Subroutine Library (OSL) Guide and Reference*, Release 2.
40. Wielinga, B.J., Schreiber, A.Th. and Breuker J.A. (1992) KADS: A modelling approach to knowledge engineering, *Knowledge Acquisition* **4**, 5-53.
41. Fourer, R., Gay, D.M. and Kernighan, B.W. (1993) *AMPL - A Modeling Language for Mathematical Programming*, Boyd & Fraser.
42. Dahal, K.P., and McDonald, J.R. (1998) Generational and steady state genetic algorithms for generator maintenance scheduling problems, in G.D. Smith,

N.C. Steele and R. Albrecht (eds.), *Artificial Neural Nets and Genetic Algorithms, Proceedings of Third International Conference in Norwich (ICANNGA'97)*, Springer-Verlag, Vienna, 260-264.

43. Dahal, K.P., and McDonald, J.R. (1997) A review of generator maintenance scheduling using artificial intelligence techniques, *Proceedings of 32nd Universities Power Engineering Conference (UPEC'97)*, 787-790.
44. Dahal, K.P. and McDonald, J.R. (1997) Generator maintenance scheduling of electric power systems using genetic algorithms with integer representation, *Proceedings of 2nd International Conference on Genetic Algorithms in Engineering Systems: Innovations and Applications (GALESIA'97)*, IEE Conference Publication 446, 456-461.
45. Dopazo, J.F. and Merrill, H.M. (1975) Optimal generator maintenance scheduling using integer programming, *IEEE Transactions on Power Apparatus and Systems* **94**, 1537-1545.
46. Egan, G.T., Dillon, T.S. and Morsztyn, K. (1976) An experimental method of determination of optimal maintenance schedules in power systems using branch-and-bound technique, *IEEE Transactions on Systems, Man and Cybernetics* **6**, 538-547.
47. Yamayee, Z. and Sidenblad, K. (1983) A computationally efficient optimal maintenance scheduling method, *IEEE Transactions on Power Apparatus and Systems* **102**, 330-338.
48. Yang, S. (1994) Maintenance scheduling of generating units in a power system, in X. Wang and J.R. McDonald (eds.), *Modern Power System Planning*, McGraw-Hill, London, 247-307.
49. Bretthauer, G., Gamaleja, T., Handschin, E., Neumann U. and Hoffmann, W. (1998) Integrated maintenance scheduling system for electrical energy systems, *IEEE Transactions on Power Delivery* **13**, 655-660.
50. Burke, K.B., Clarke, J.A. and Smith, A.J. (1998) Four methods for maintenance scheduling, in G.D. Smith, N.C. Steele and R. Albrecht (eds.), *Artificial Neural Nets and Genetic Algorithms, Proceedings of Third International Conference in Norwich (ICANNGA'97)*, Springer-Verlag, Vienna, 265-270.

Chapter 5

TRANSMISSION NETWORK PLANING USING GENETIC ALGORITHMS

M. R. Irving, H. M. Chebbo And S. O. Orero
Department of Electrical Engineering and Electronics
Brunel University, Uxbridge
UB8 3PH, UK

5.1 Introduction

There has recently been considerable interest in Genetic Algorithms, and other evolutionary techniques, among researchers in various application fields. The simplicity, flexibility and robustness of such algorithms has opened up new areas of application, and has also encouraged a re-appraisal of some traditional problems that were either very difficult or even intractable for traditional optimisation techniques.

This chapter presents an approach to Genetic Algorithms (GAs) based on a stage by stage development from very simple random search methods. Some of the advantages of GAs are highlighted, and their disadvantages are also considered. Recent developments, designed to overcome some of the disadvantages, are presented.

A significant application of GAs in the power systems field is explored. This is the problem of power system network planning. A new problem formulation is introduced which seeks to identify the optimal network design over an extended time horizon given an expected pattern of power demand and generation. A simplified model of the transmission network requirements of England and Wales is used for computational test purposes. It is shown that the GA can produce optimal results for certain classes of problem in which a theoretical optimal solution is known, and that for more general problems, where no theoretical optimal solution is available, the GA produces good solutions.

5.2 Genetic Algorithms

5.2.1 BACKGROUND

Genetic Algorithms [1,2] are computational models of biological evolution. There are many ways of defining and interpreting GAs, but for the purposes of this chapter we will primarily consider them as computational optimisation techniques. As such, the GAs themselves must compete with alternative heuristic methods such as Simulated Annealing, Tabu Search, Ant-Colony Methods etc., and also with mathematical programming techniques such as non-linear programming. Since their introduction GAs have proved to be a great success, with a large number of researchers and practitioners adopting them to solve a wide variety of problems.

5.2.2 DEVELOPMENT OF GAs AND SOME RELATED ALGORITHMS

A computational optimisation technique is concerned with selecting the best solution from a set of possible solutions. We may represent a possible solution as a string:

$$\mathbf{s} = (s_1, s_2, \ldots, s_n)$$

Where s_i is a symbol from some alphabet

The symbols used are often binary digits, in which case the string is a binary string.

Initially, we will consider unconstrained problems. In other words, any possible string that can be constructed from the alphabet is assumed to represent a feasible solution. In order for there to be a 'best' or optimal solution there must be a well-defined formula or algorithm which computes a cost function C(s) for any possible string. An optimal string is then a string s* for which:

$$C(s^*) = \min_{\text{for any } s} C(s)$$

In modeling an optimisation problem there may be a variety of ways to define a suitable string, and alternative alphabets may be available, any of which adequately represent all the possible solutions. The mapping between the string and the possible solutions of the problem is referred to as 'coding' and is a crucial aspect of the application of GAs. Generally, an efficient coding should take into account the following points:

- Redundancy of the code should be as low as possible, i.e. the string length and the number of possible alphabet symbols (the cardinality of the alphabet) should be as low as possible, consistent with an adequate representation of the problem.
- The de-coding process should be straightforward, allowing rapid calculation of the cost of any candidate string.

- Problem constraints should be represented implicitly by the coding (where possible), so that the majority of possible strings do represent feasible candidate solutions.
- The individual symbols in the string should be as independent as possible in terms of their effect on the overall cost. Highly interactive codes, in which individual symbols are only meaningful in combination with many other symbols, will impede the GA's search for an optimal solution.

Having defined an alphabet, a string and a cost function, it is possible to introduce a simple heuristic search method. This technique is often referred to as 'Monte Carlo Search' or 'Trial-and-Error Search'.

5.2.2.1 Monte Carlo Search

This algorithm maintains a single 'incumbent' candidate solution that is repeatedly subjected to random changes. Any change that produces an improvement (a reduction in cost) is retained. In GA terminology the random changes can be called 'mutations'. The algorithm can be described as follows:

1. Select an initial string s^0 at random
2. Evaluate the cost $C^0 = C(s^0)$
3. Select a position j in the string at random
4. Change the symbol at position j to a different random symbol. This creates a new string s^i
5. Evaluate the cost of the new string
6. $C^i = C(s^i)$
7. if $C^i <= C^{i-1}$ then
     s^i is accepted as the new incumbent
   else
     s^i is rejected and s^{i-1} is retained as the incumbent
   endif
8. Re-iterate from 3. until no improvement has occurred for a number of iterations.

This is a very simple algorithm, and it has an obvious drawback. In fact, the algorithm will never 'escape' from a local minimum. When the incumbent string is a local optimum solution, by definition there are no *single* symbol changes that produce an improvement. At this point we may note that some alternative methods such as Simulated Annealing [3] have been introduced which allow this type of algorithm sometimes to proceed 'up hill' allowing an escape from a local minimum.

In an attempt to seek a global optimum, it would be possible to re-start the above algorithm from a number of different random starting strings, and these various iterative sequences could of course be processed in parallel. In GA terms we would then have a population of candidate solutions. Although this algorithm would be ideally suited to a parallel processing computer there is no information exchange among the population, and the results obtained would be identical to those obtained by the equivalent sequential approach.

In the parallel algorithm, many population members could be expected to perform poorly with respect to other population members (e.g. they may be trapped in a high-cost local minimum). One approach towards overcoming this difficulty would be to consider a 'survival of the fittest' strategy by introducing the concept of 'generations' of the population, and selecting the population members to proceed from one generation to the next. Such concepts define a simple 'evolutionary algorithm'.

5.2.2.2 Simple Evolutionary Algorithm

A basic evolutionary algorithm can be described as follows.

1. Select (at random) an initial population of strings $\{ s^0_1, \ldots, s^0_k, \ldots, s^0_P \}$, where P is the number of population members.
2. Evaluate the cost $C^i_k = C(s^i_k)$ for all k.
3. Perform random mutations on existing population members to produce new population members.
4. Select a subset of population members to be deleted. The selection should depend on the cost of each population member, so that the individuals with high cost are more likely to be deleted.
5. Re-iterate from 2. until no improvement in the best member of the population has occurred for some iterations.

An important component of the above algorithm is the choice of the subset of population members to delete (or conversely the selection of the individuals to survive into the next generation). This process is called 'selection' and a variety of techniques are possible. An obvious approach would be to simply retain the subset (e.g. half of the population) with the lowest costs. However, as well as retaining the 'good' candidates we may wish to randomly allow some of the higher cost individuals to pass through into the next generation in an attempt to escape local minima. To increase the chance of finding the global minimum we would also wish to preserve 'diversity' in the population (i.e. avoid too much duplication of strings or sub-strings). Consequently, techniques such as 'roulette wheel selection' and 'tournament selection' have been developed. In tournament selection some (usually small) subset of the population is chosen (usually at random) and the higher cost members of the subset are deleted. In roulette wheel selection a fitness measure is defined (as a monotonic function of cost) which determines the likelihood of an individual being selected for survival. By analogy with a physical roulette wheel, a random experiment is performed with the probability of selecting any individual being proportional to its fitness, which is inversely related to cost. The mapping from cost value to fitness value, known as fitness scaling, is designed to give a *reasonable* likelihood of survival for any string even in circumstances where cost function values can exhibit extremely high, or extremely low, numerical variation. An advantage of tournament selection is that fitness scaling is not needed.

Often 'elitism' is introduced in the selection mechanism, whereby the best individual or the best few individuals are automatically retained, so that each generation does at least include the individual which is 'best so far'.

5.2.2.3 Genetic Algorithm

The simple evolutionary algorithm, described previously, requires one further refinement to be regarded as a Genetic Algorithm. This is the 'crossover' operation. Crossover allows information to be exchanged between individuals in the population. Two 'parent' strings are selected (usually at random) and a new 'child' string is created by combining random sub-strings from the parents. There are various alternative techniques such as single-point, multi-point and uniform crossover. Uniform crossover is now generally regarded as the best option, and will be considered here. Taking as an example two parents par_0 and par_1 with string length 12 and using binary symbols:

$$par_0 = (\ 101100010110\ )$$
$$par_1 = (\ 001010111010\ )$$

define a (random) 'mask' string,

$$m = (\ 101100010110\ )$$

then, according to uniform crossover, a child is defined as containing the corresponding symbol from parent 0 for every bit position in the mask with value 0, and containing the corresponding symbol from parent 1 for every bit position in the mask with value 1, giving:

$$ch_1 = (\ 001000010010\ )$$

A second child can also be obtained by using the bit-wise inverse of the same mask:

$$ch_2 = (\,101110111110\,)$$

The conventional definition of crossover probability is the probability of a mask bit being given a value of 1. With this definition, a probability of 0.5 means that a child has an equal likelihood of inheriting bit values from either parent, a probability of 1.0 means that child 1 is identical to parent 1, etc.

An alternative definition can also be considered, where the crossover probability is defined by the following simple rules:

1. Randomly choose the first bit of the mask string as either 0 or 1, with equal probability.
2. Randomly choose each consecutive bit of the mask string with a given probability of being the same as the preceding bit.

Using this definition, a crossover probability of 0.5 gives statistical properties which are identical to those of the previous definition, but other crossover probability values bias the 'expected run length' of consecutive 1s or 0s in the mask. This definition is therefore more consistent with the principles of single-point and multi-point crossover. We have found this definition to be more useful for the problems we have studied.

We can now refine the previous algorithm to produce a 'canonical' GA as follows:

1. Select (at random) an initial population of strings $\{\,s^0_1,\ldots,s^0_k,\ldots,s^0_P\,\}$, where P is the number of population members.
2. Evaluate the cost $C^i_k = C(s^i_k)$ for all k.
3. Perform random crossovers, introducing new population members.
4. Perform random mutations.
5. Select a subset of population members to be deleted. The selection should depend on the cost of each population member, so that high-cost strings are more likely to be deleted.
6. Re-iterate from 2. until no improvement in the best member of the population has occurred for some iterations.

The use of crossover is the defining characteristic of a GA and has a very powerful influence on the speed of convergence of the population. To appreciate this, it is useful to consider the concept of a 'schema'. If the two parents given in the above example are considered, it is apparent that the bit values in some positions are common to both parents. This can be expressed by stating that both parent 0 and parent 1 satisfy the following schema:

(*01**0*1**10)

the character * in the schema is a 'wild-card' meaning that a bit in that position can be either 1or 0.

Of course, each parent also satisfies a very large number of other schemas, which are not in common with the other parent. For example, parent 0 also satisfies the following schemas: (1011********), (****000*****), and (*******10*10).

The schema theory [1,2,4] provides a formal mathematical framework for the analysis of GAs. In general terms, each parent may satisfy some schemas which are associated with relatively high fitness, indeed some schemas may be in common with an optimal (or a local-optimal) solution. Such schemas could loosely be called 'good' schemas. When two parent strings are combined via crossover, there is a chance that the child will satisfy good schemas of both parents, and the child will then usually be 'fitter' (lower cost) than either parent. In this case a new 'good' schema with fewer 'wild-cards' has been introduced into the population. Naturally, it is also possible that the child will have lower fitness than either parent, but then the child is less likely to be selected for the next generation. The combined effect of crossover and selection pressure is that good schemas emerging in the population tend to be strongly reproductive *in their own right*. As the generations proceed, good schemas rapidly propagate throughout the population, and tend to combine via crossover to form new good schemas having fewer wild-cards. Within a few generations, duplicates of a few distinct strings may dominate the population. At this stage the dominant string (or strings) may still be sub-optimal or only local-optimal, and in this case the GA is said to have exhibited premature convergence. The only factor preventing this in the canonical GA is the randomising effect of mutations. In a later section, we will describe a 'niching' method that seeks to preserve population diversity to counteract this problem.

5.2.3 ADVANTAGES OF GENETIC ALGORITHMS

Some of the advantages of GAs, for the purposes of computational optimisation, are discussed in the following sub-sections.

5.2.3.1 Global Optimisation

Many practical optimisation problems contain multiple local optima. An important advantage of GAs is the possibility that more than one local optimum will be explored and there is a chance that the GA *may* discover a global optimal solution. This property is shared with many other heuristic search techniques (such as Simulated Annealing, Tabu Search, Ant-Colony Search etc.) whereas mathematical programming techniques are generally equipped to seek only a local optimum. Of course, mathematical programming techniques can be adapted to search for a global optimum; for example by using a large number of randomly selected starting points. The relative merits of the GA approach and modified mathematical programming approaches have not yet received sufficient attention in the literature. There is also considerable scope for further investigation of hybrid methods intending to combine the advantages of GAs with conventional mathematical programming techniques.

5.2.3.2 Generality of Objective Functions

The only restriction on the type of objective function that can be accommodated in a GA is simply that it must be a computable scalar function of any feasible candidate solution. This is in contrast to most mathematical programming based techniques which may impose quite severe restrictions (such as linearity, differentiability, continuity, convexity, etc.) on the type of objective function which can be accepted.

5.2.3.3 Relative Ease of Programming

Genetic Algorithms are relatively straightforward to program, in contrast with mathematical programming techniques which generally require sophisticated linear algebra routines, calculation of partial derivatives, robust line searching methods, etc.

5.2.3.4 Algorithm Flexibility

There are many possible variants to Genetic Algorithms, some of which will be more efficient for particular problems than others. In this chapter, a particular GA variant, introduced by Mahfoud [5], and known as Deterministic Crowding, is developed. This approach has been found to be particularly efficient for problems such as generator scheduling [6,7], hydro-thermal power system co-ordination [8] and transmission network planning [9].

5.2.3.5 Numerical Robustness

The simplicity of GAs leads to a further advantage. This is that they do not suffer from the numerical robustness problems which can occur in some mathematical optimisation techniques (e.g. matrices becoming ill-conditioned or singular, iteration steps becoming too short prematurely, etc.)

5.2.4 DISADVANTAGES OF GENETIC ALGORITHMS

Despite the considerable advantages of GAs there are, of course, some disadvantages that must be considered:

5.2.4.1 Potential Premature Convergence

As outlined above, crossover and selection tend to cause unstable reproduction of schemas associated with above-average fitness, which can result in rapid loss of diversity in the population and premature convergence to a sub-optimal solution.

5.2.4.2 Long Computing Times

The number of candidates considered in a GA population and the number of generations which must be simulated before convergence is obtained, can combine to produce excessive run times for problems in which the fitness calculation is non-trivial.

5.2.4.3 Parameter Tuning

Many GA variants include a variety of parameters that can be adjusted by the user (e.g. mutation and crossover probabilities, population size, selection mechanisms etc.). In some ways the ability to 'tune' these parameters can be seen as an advantage, because the analyst can adjust the GA to suit his particular class of problems. However, since this tuning is generally based on extensive numerical experiments and can require considerable effort, methods that require less tuning are often preferred.

5.2.4.4 Modeling Constrained Problems

GAs are probably better suited to unconstrained optimisation problems than to constrained problems. In some cases a special coding technique can be defined so that any candidate string is guaranteed to satisfy all (or some of) the constraints. This is a very advantageous approach, when it is available. In other cases penalty factors can be applied to convert a constrained problem into an unconstrained problem. The simplistic approach, whereby any candidate is checked against the constraint set and is rejected outright if found to be infeasible, can be very inefficient if a large proportion of the candidates that are generated prove to be infeasible.

5.2.4.5 Non-deterministic Solutions

Most mathematical programming techniques provide additional information (such as gradient values, approximate Hessian matrix values, duality- or complimentarity-gap values, etc.) which can provide some reassurance that a good solution has been obtained. The results obtained by GAs, or by other similar probabilistic search methods, do not incorporate this type of subsidiary information. This can be an important drawback for significant technical and commercial applications where a high level of quality assurance is required. In some respects it can be regarded as an advantage that a better solution might be obtained from a GA simply by allowing further evolution, greater population sizes etc., but the difficulty is that for non-trivial problems some uncertainty always remains as to the quality of the final solution produced.

5.2.5 DETERMINISTIC CROWDING GENETIC ALGORITHM

An area of further enhancement to the standard GA has been the introduction of niche methods [2,5,10]. These methods reduce competition among population members when there is a sufficiently large difference (or distance) between them, allowing sub-populations centering on good solutions (niches) to co-exist. Niche methods fall into two broad categories: 'crowding' and 'sharing'. Crowding methods restrict the replacement of individuals by discouraging competition among widely differing individuals, while sharing methods de-rate an individual's effective fitness when similar individuals exist. The potential disadvantage of most niche-based methods is the computational burden of comparing each individual to many other individuals, in order to evaluate the similarity measure.

Niche GA models are inspired by a corresponding natural ecological phenomenon, where similar members of a natural population compete for the same resources. A niche GA attempts to maintain a population of diverse individuals in the course of the simulated evolution. Genetic algorithms that incorporate these ideas are thus better able to locate multiple local-optimal solutions within a single population. The particular niche method described here is based on the crowding concept.

In a crowding GA the new population is created by allowing child strings to replace the parents that are most similar to themselves. De Jong [10] proposed a crowding factor model where only a fraction of the population reproduces and dies in each generation, with each newly created population member replacing an existing member, preferably the most similar. Following analysis and modification of De Jong's (and other) crowding methods, Mahfoud [5] proposed the Deterministic Crowding Genetic Algorithm (DCGA). Mahfoud's algorithm is computationally efficient, since each offspring is only compared with its two parents, and competes only with the more similar parent. The DCGA approach provides selection pressure *within* but not *across* regions of the search space, leaving the search across regions to the crossover operator.

The main properties of DCGA that distinguish it from other GA models are:
- selection and population replacement are combined into a single step,
- random crossover is performed on the whole population,
- there is no need for any scaling mechanism, and
- inferior parents are replaced by their more similar children.

The DCGA randomly pairs all population members in each generation to yield P/2 pairs of parents, for a population of size P. Each pair undergoes crossover, possibly followed by mutation, to produce two offspring. Each of the two offspring competes with one of the two parents, chosen according to a similarity measure. The fitter among them forms the population of the next generation. For example, given a pair of parents and their two offspring, two sets of parent-offspring tournaments are possible:

set 1
- parent 1 against child 1
- parent 2 against child 2

set 2
- parent 1 against child 2
- parent 2 against child 1

The set of tournaments which forces competition among more similar individuals is held, where similarity is defined as the average distance between the parent-child combinations in the set.

The DCGA can therefore be summarised as follows:

1. Select (at random) an initial population of strings $\{ s^0_1, \ldots, s^0_k, \ldots, s^0_P \}$, where P is the number of population members.
2. Evaluate the cost $C^i_k = C(s^i_k)$ for all k.
3. While (i <= P)
4. Choose two parents par_1 and par_2 at random, without replacement.
5. Perform crossover and mutation to produce offspring ch_1 and ch_2.
6. Evaluate fitness (f) of parents and offspring.
7. If ($dist(par_1,ch_1) + dist(par_2,ch_2) <= dist(par_1,ch_2) + dist(par_2,ch_1)$) then
       if ($f(ch_1) > f(par_1)$) replace par_1 by ch_1
       if ($f(ch_2) > f(par_2)$) replace par_2 by ch_2
   else

             if ($f(ch_2) > f(par_1)$) replace par_1 by ch_2
             if ($f(ch_1) > f(par_2)$) replace par_2 by ch_1
       endif
8. Next i
9. Re-iterate from 2. until no improvement in the best member of the population has occurred for some iterations.

The method of choosing the two parents *without replacement* is important, since it inherently gives the DCGA the property of elitism. Every population member (including the 'best-so-far') *must* enter a tournament as a parent, and can only be eliminated from the population if it is to be replaced by a fitter child.

5.2.6 SOLUTION OF CONSTRAINED PROBLEMS

Simple constraints, such as upper and lower limits on variables, can easily be represented implicitly within the problem coding. In some cases, it may also be possible to represent more complex constraints by an appropriate choice of coding. However, for general constraints, it is usual to represent constraints via penalty functions added to the cost function. An *overall cost function*, consisting of the original cost function *plus* additional penalty costs for any violated constraints, is then optimised. A practical example of the use of penalty functions is given in a later section of this chapter. In general, penalty functions should be chosen to satisfy the following requirements.

- The penalty function should be progressive, so that a more severe violation of a constraint attracts a higher penalty cost - allowing the GA to be 'guided' towards feasibility.
- The penalty factors for each violated constraint should be summed to form the overall penalty cost, so that a well-behaved penalty cost surface is produced.
- The value of penalty costs should be higher than actual costs, so that a solution at the optimum of the overall cost function (including any penalty costs) would not include any non-zero penalties (i.e. violated constraints). If the penalty cost values are too small relative to the actual costs, it would be possible for the GA to 'trade-off' some constraint violations against cheaper actual costs, possibly arriving at an overall 'optimum solution' which includes some violated constraints.
- On the other hand, the value of penalty costs should not be too high in relation to the actual costs, so that the overall cost surface (including penalty costs) is reasonably well-conditioned. In this respect it is frequently possible to determine a realistic penalty cost value in economic or physical terms. For example, in some problem formulations the maximum available level of a commodity may be represented as a constraint, whereas in reality some additional supply may actually be available (albeit with a higher cost), e.g.

from a spot-market, or by using emergency resources, etc. In such cases a suitable penalty cost value for the constraint is already available.

It is well known in conventional mathematical programming that the use of penalty costs can lead to an ill-conditioned overall problem. This arises from the need to satisfy both of the final two points mentioned above. By applying penalty values that are large enough to guarantee feasible solutions it is difficult to avoid creating severely distorted overall cost function surfaces, which are then problematic for a mathematical programming method. This drawback of the penalty factor approach is much less significant for GAs, and similar methods, since they do not impose any particular requirements on the mathematical properties of overall cost function surface. The fact that an overall cost function (including penalty costs) may not be differentiable or smooth, and may have poor curvature properties (e.g. having singular or ill-conditioned Hessian matrices at certain points) would not necessarily impede the progress of a GA.

5.3 The Transmission Network Planning Problem

The planning problem that will be considered here is the design of an electricity transmission network, which is as economical as possible while providing a reliable energy supply. The mathematical formulation of the problem leads to a complicated, integer-valued, non-convex, non-linear mathematical programming problem. The complexity of the problem arises mainly from the large number of problem variables, combined with the variety of economic factors and technical constraints that must be taken into account. Several classical optimisation techniques have been applied to various formulations of the problem; including linear programming [11], non-linear programming [12], mixed integer programming [13], Bender's decomposition [14,15] and other techniques [16,17,18]. Some heuristic methods have also been used, often based on sensitivity analysis [14,18].

The attributes of the mathematical formulation of the problem seem well suited to the application of GAs, and in the present chapter a Deterministic Crowding GA will be applied to its solution.

5.3.1 PROBLEM DESCRIPTION

For good practical reasons, transmission planning engineers have generally taken an incremental approach, and have tended to evaluate a relatively small number of alternative expansion plans over a relatively short time horizon.

The general form of the network expansion problem can be stated as follows:

Given:

(i) the load and generation patterns for the target year,

(ii) an existing network configuration,

(iii) all possible new routes (particularly their location and length)

(iv) the available transmission line types and their corresponding costs.

Determine:

an optimum network configuration which supplies the loads with the energy required at the lowest possible cost, while meeting specified transmission security standards.

An alternative approach, which could be referred to as a 'non-incremental', or a 'green-field' approach, is to design an optimal network based on available routes but without regard to the existing configuration of transmission lines. This problem can be stated as:

Given:

(i) the load and generation patterns for the target year,

(iii) all possible routes (particularly their location and length)

(iv) the available transmission line types and their corresponding costs.

Determine:

an optimum network configuration which supplies the loads with the energy required at the lowest possible cost, while meeting specified transmission security standards.

This non-incremental approach is impractical for monthly and annual planning decisions, which must necessarily be compatible with the existing network configuration. Nevertheless, there is some motivation for developing the green-field approach as a complementary facility. In particular, for de-regulated utilities there may be a need for an optimal network design against which the existing configuration may be benchmarked. Furthermore, it is useful for planners (who are routinely applying the practical incremental approach) to have a long-term future target available for reference. For example, if an optimal green-field network design is available, engineers can gain insight into which of their plans are heading 'towards' the optimal design. Conversely some short-term plans may appear to be diverging from the future optimum, and could therefore be regarded as less effective in the longer term. The DCGA model has therefore been applied to the non-incremental transmission network planning problem.

5.3.2 GENETIC ALGORITHM MODEL

To adopt the GA approach, the problem must be represented in a suitable format that allows the application of the GA operators. This is the problem encoding process. As described above, the DCGA maximises a single variable, the fitness function. Hence, the cost function and constraints of the network planning problem must be transformed into an appropriate fitness measure.

5.3.2.1 Problem Encoding

Binary representation has been widely used for GA coding, partly because of the ease of binary manipulation, and also since much of the theory of GAs is based on the use of the binary alphabet. However, some recent GA implementations have used more general symbol alphabets. We have conducted some limited experiments to assess the relative merits of a special symbol alphabet for the present problem, but concluded that the binary approach was slightly superior in the cases considered.

Adopting a binary representation, each member of the population corresponds to string of 1s and 0s, and represents a given transmission network design. Every route is assigned a binary number, which represents the type of transmission line selected for that route. In the examples presented later in this chapter, 14 different line types are available. These include line type 0 to represent an unused route, and various single-circuit and double-circuit line types. If it is anticipated that two (or more) double circuits may be required on a given route, it is necessary to specify two (or more) parallel routes in the problem description. Associated with each line type are the corresponding capital cost, electrical resistance (for calculating energy loss), etc. A bit string of length 4 is needed to represent the 14 different possible line types for each route. This implies that two undefined strings, which do not have any corresponding line type, may occur. If these strings are generated during the execution of the GA, the GA operator that produced the invalid string is repeated until a valid string is produced. The 4-bit strings for each route are concatenated to form an overall string which represents a particular network design, and which is a member of the GA population. In the example problems here, there are 49 available routes, leading to an overall string length of 196-bits (i.e. 49x4).

5.3.2.2 Fitness Function

The objective of transmission planning is to minimise annuitised investment cost together with the cost of energy losses, while satisfying system constraints. Any violation of constraints implied by a candidate solution is handled via the penalty function approach, in which penalty costs are incorporated into the fitness function so as to reduce the apparent fitness of any infeasible candidate.

The overall objective function considered is therefore as follows:

$$F = \text{lecf} \cdot \Sigma \, lg_i \cdot llf_i + \text{lccf} \cdot \Sigma \, lg_i$$

$$+ E \cdot \Sigma \, eg_i + \Psi \cdot \Sigma \, |pf_i|$$

$$+ \Phi \Sigma \, (\,|pf_i| - ratl_i\,) + \Sigma \, l_i \cdot costl_i$$

$$+ p_0 \cdot (\,nisl - 1\,) + \Sigma \, \mu_i \cdot \Sigma \, (\,|pf_k| - ratl_k\,)$$

where:

F	is the overall fitness value,
	(Σ implies summation over the appropriate elements),
lecf	is the loss energy cost factor,
lg_i	are the thermal energy losses for each line,
llf_i	are the loss load factors for each line,
lccf	is the energy loss capacity cost factor,
E	is environmental impact cost factor,
eg_i	are the environmental impact factor for each line,
Ψ	is a penalty cost factor for unsatisfied loads (based on the power flow that would be required to satisfy the load),
pf_i	are the power flows on each line,
Φ	is a penalty cost factor for line overloading,
$ratl_i$	are the power flow ratings of each line,
l_i	are the lengths of each line,
$costl_i$	are the annuitised capital costs (per unit length) of each line,
p_0	is a penalty cost factor for network islanding,
nisl	is the number of network islands,
μ_i	is a penalty cost for line overloading following an outage.

Any additional factors that are pertinent to the planning problem can easily be added to the fitness function, provided that they can be computed as a function of a candidate string. However, the computer time required to evaluate the fitness function will have a direct impact on the overall solution time. To allow more rapid evaluation of the present network planning model, the DC load-flow approximation [19] has therefore been adopted. The DC approach provides a linear active power flow model that is sufficiently accurate for the present application, and can be used efficiently in conjunction with the Householder modified matrix formula for outage studies.

The candidate string generated by the GA is interpreted as a specific power network configuration. Power flows, overloads (if any), and approximate energy losses are then evaluated using the DC model. Further (modified) DC models are then constructed for every outage case to be considered. In order to meet security requirements, it is necessary that no line shall be overloaded when one or more circuits are removed from the planned intact network. In the examples presented here, the so-called 'N-line' security analysis is adopted, where the optimum network plan is designed to withstand any single line outage without overloading. Furthermore the possibility of 'islanding', whereby the network operates as two or more disconnected parts, is also precluded (via high penalties) both in the intact network and in any of the outage case networks.

5.4 Experimental Results

5.4.1 TEST PROBLEMS

The DCGA technique has been applied to a range of problems derived from a 23-node 49-route transmission network design problem that represents a simplified version of the England and Wales transmission network. The demand and generation profiles represent predicted peak values for the near-term future. The GA has the ability to choose a line type in the range 0 to 14 for every available route, with type 0 representing an unused route and types 1 to 14 representing a variety of standard transmission line operating at voltage levels 275kV, 400kV or 750kV. The highest voltage level (750kV) is not used in the UK at present and thus represents a hypothetical option.

The various classes of problem are described in following sections. For two of the problem classes the nature of the optimal solution is known in advance from theoretical considerations. These cases provide an opportunity to test the validity of the solutions proposed by the GA approach. The remaining two problem classes are more realistic, but are also more complex and do not have known optima. For these problems it is not possible to assess the solutions provided by GAs theoretically and it is necessary to rely on numerical experiments. Further research is also underway which will apply an alternative solution approach based on ant-colony optimisation to generate comparative results for these cases.

5.4.1.1 Problem Class A: Cost of Energy Losses Only

For this problem class, the cost of energy losses is to be minimised subject only to the satisfaction of required generation and load levels throughout the network. Other costs and constraints are temporarily neglected. From considerations based on simple electrical network theory, it is apparent that an optimal solution for this case consists of a transmission network design with each available route occupied by the line type having the lowest electrical resistance.

5.4.1.2 Problem Class B: Investment Cost Only

In this problem class, only the annuitised capital cost of transmission lines is considered, subject to satisfying required generation and load levels. Other costs and constraints are temporarily neglected. A theoretical optimal solution for this problem class is known to consist of a network design based on a minimum length spanning tree (i.e. a radial network with shortest possible total line length) in which the line type with lowest capital cost is used throughout.

5.4.1.3 Problem Class C: Cost of Energy Losses and Investment Cost

This problem class represents a realistic planning problem in which all factors, except security against outages, are considered. In particular, the solutions obtained for this class of problems, show how energy loss costs are to be 'traded-off' against initial capital costs. There are no known optimal solutions for this class of problems, and so the solutions obtained by applying GAs are of significant interest.

5.4.1.4 Problem Class D: Cost of Energy Losses, Investment Cost and Security Analysis

This final problem class represents the full-scale problem with all factors considered. Again, no optimal solutions are known in advance. The comparison of the solutions obtained for problem class C with those obtained for class D allow the additional cost of designing a secure network to be assessed.

5.4.2 TEST RESULTS

Several tests have been performed on the sample problem, and considering the four classes of cost function outlined above. The GA parameter settings and the results obtained are summarised in Table 1.

Table 1: GA parameters settings and optimisation results

Problem Class A (Loss Cost Minimisation)	Population Size 500 Mutation Probability 0.009 Crossover Probability 0.5 Number of generations 900	Loss Cost £ 238,111,700 Total Cost £ 238,111,700
Problem Class B (Investment Cost Minimisation)	Population Size 500 Mutation Probability 0.009 Crossover Probability 0.1 Number of gens. 1600	Inv. Cost £ 422,362,000 Total Cost £ 422,362,000
Problem Class C (Full formulation but without consideration of security constrainits)	Population Size 500 Mutation Probability 0.009 Crossover Probability 0.5 Number of gens. 3400	Inv. Cost £1,227,724,000 Loss Cost £ 715,025,100 Total Cost £1,942,749,100
Problem Class D (Full problem formulation)	Population Size 500 Mutation Probability 0.007 Crossover Probability 0.09 Number of gens. 1300	Inv. Cost £1,493,527,000 Loss Cost £ 768,675,000 Total Cost £2,262,202,000

The results shown in Table 1 represent the best solutions that have been obtained, and illustrate that the GA parameters required some tuning to obtain the best result for each problem class. The solutions obtained for problem class A and problem class B agreed with the known theoretical optima, giving the users a degree of confidence in the GA approach. However, it may be noted that the number of generations required to converge was generally greater for the more realistic problems, suggesting that these problems may be harder to solve. Further work is needed to fully investigate the quality of the best solutions obtained for these realistic problems, but the solutions did appear to be of high quality when inspected manually by experts.

5.5 Future Work

Further refinements can be introduced into the planning model, as a consequence of the flexibility of the GA approach. Work is underway to include representations of network maintainability and the additional cost of the transformers needed to connect lines of differing voltage levels. Further work is also in hand to investigate the quality of the best results obtained by comparison with solutions obtained using other optimisation techniques.

5.6 Conclusions

The transmission planning problem is a difficult non-linear, non-convex, discrete-variable constrained optimisation problem. Genetic Algorithms, and in particular Deterministic Crowding, have been shown to be highly suitable for application to this problem. For certain classes of problem, where a known theoretical optimal solution exists, GAs were found to be capable of locating the exact optimum. More general classes of problems have also been studied, and in these cases although no optima is known it is believed that the solutions produced by the GA are of high quality. It is concluded that GAs offer a very useful new approach for the solution of power system network planning problems.

5.7 Acknowledgment

The authors would like to acknowledge the financial support of EPSRC and the National Grid Company (for Mrs. H. M. Chebbo) and of the British Council (for Dr. S. O. Orero). They would also wish to thank Dr. L. Dale, Dr. M. Zhu and Ms. U. Bryan of the National Grid Company for providing the definition of the transmission network planning problem and for many helpful discussions during the research investigation.

5.8 References

1. Holland, J.H. (1975) Adaptation in Natural and Artificial Systems, *MIT Press*, Cambridge, Mass.
2. Goldberg, D.E. (1989) Genetic Algorithms in Search, Optimisation and Machine Learning, *Addison-Wesley*, Reading, Mass.
3. Kirkpatrick, S., Gellat, C.D. and Vecchi, M.P. (1983) Optimization by Simulated Annealing, *Science*, 220, pp 671-680
4. Whitley, D. (1994) A Genetic Algorithm Tutorial, *Statistics and Computing*, 4, 2, pp 65-85
5. Mahfoud, S. (1992) Crowding and Pre-selection in Genetic Algorithms, from *Parallel Problem Solving from Nature 2, R. Manner and B. Manderick (Ed.s)*, Elsevier, pp 27-36.
6. Orero, S.O. (1996) Power Systems Generation Scheduling and Optimisation using Evolutionary Computation Techniques, Ph.D. thesis, *Brunel University*, Middlesex, UK
7. Orero, S.O. and Irving, M.R. (1997) Large Scale Unit Commitment using a Hybrid Genetic Algorithm, *Int. Jnl. of Electrical Power and Energy Systems*, 19, pp 45-55.
8. Orero, S.O. and Irving, M.R. (1998) A Genetic Algorithm Modelling Framework and Solution Technique for Short Term Optimal Hydro-thermal Scheduling, *IEEE Trans. Power Systems*, 13, 2, pp 501-518
9. Chebbo, H.M. and Irving, M.R. (1997) Application of Genetic Algorithms to Transmission Planning, *IEE GALESIA conf.*, Glasgow, pp 388-393
10. De Jong, K.A. (1975) An Analysis of the Behaviour of a Class of Genetic Adaptive Systems, Ph.D. thesis, *University of Michigan*, Ann Arbor, USA

11. Chanda, R.S. and Bhattacharjee, P.K. (1994) Application of Computer Software in Transmission Expansion Planning using Variable Load Structure, *Electric Power Systems Research*, 31, pp 13-20
12. Giles, M.L. (1986) Optimum HVAC Transmission Planning – a New Formulation, *IEEE Trans. Power Systems*, 1, pp 48-56.
13. Santos, A., Franca, P.M. and Said, A. (1989) An Optimisation Model for Long Range Transmission Expansion Planning, *IEEE Trans. Power Systems*, 4, pp 94-101.
14. Pereira, M.V.F., Pinto, L.M.V.G., Cunha, S.H.F. and Oliveira, C.G. (1985) A Decomposition Approach to Automated Generation / Transmission Expansion, *IEEE Trans. Power Apparatus and Systems*, 104, pp 3074-3083.
15. Romero, R. and Monticelli, A. (1994) A Zero-One Implicit Enumeration Method for Optimising Invetsments in Transmission Expansion Planning, *IEEE Trans. Power Systems*, 9, 3, pp 1385-1391.
16. Rudnick, H., Palma, R., Cura, E. and Silva, C. (1996) Economically Adapted Transmission Systems in Open Access Schemes – Application of Genetic Algorithms, *IEEE Trans. Power Systems*, 11, pp 1427-1433.
17. Monticelli, A., Santos, A., Pereira, M.V.F., Cunha, S.H.F., Parker, B.J. and Praca, J.C.G. (1982) Interactive Transmission Network Planning using a Least-Effort Criterion, *IEEE Trans. Power Apparatus and Systems*, 101, 10, pp 3919- 3925
18. Latorre-Bayonna, G. and Perez-Arraga, I. (1994) Choppin - a Heuristic Model for Long Term Transmission Expansion Planning, *IEEE PES Winter Meeting*, New York
19. Knight, U.G. (1972) Power Systems Engineering and Mathematics, *Pergamon Press*, London, UK

Chapter 6

ARTIFICIAL NEURAL NETWORKS FOR GENERATOR SCHEDULING

Michael P. Walsh And Mark J. O'Malley
University College, Dublin, Ireland

6.1 Introduction

The Hopfield neural network was first developed in 1982[10] and has since found applications in many optimisation problems. This chapter first describes the development of the Hopfield network and illustrates its applicability to optimisation problems. A selection of applications to the scheduling problem will then be discussed. The Augmented Hopfield Network is then described and its application to the generalised scheduling problem is illustrated. Finally an overview of other neural network approaches to generator scheduling is provided.

6.2 Hopfield Networks

The original Hopfield network consists of a single layer of simple processing elements known as neurons [10][11]. These neurons attempt to model the biological neurons whose interaction provide the brain with its processing power. Neurons may be classified by the transfer function which relates their input and output. In this chapter a discrete and a continuous transfer function will be defined. Hopfield networks are single layer, recurrent neural networks. The action of a Hopfield network is to minimise an energy function. The outputs of all neurons are initialised by a random number generator [19]. These outputs are then fed back via a weighted sum to provide the inputs to the neurons. Based on these inputs a new set of outputs may be computed. These outputs are then fed back, as before, yielding a new set of outputs. An iteration is said to have occurred when all neuron outputs have been updated. This process continues until the outputs do not change on successive iterations at which stage convergence is said to have occurred. It will be shown that the state converged to is a minimum of an energy function whose parameters are the connection weights.

6.2.1 DISCRETE HOPFIELD NETWORKS

In the discrete Hopfield network [10] the ith neuron has as its input (U_{di}) the weighted ($T_{di \rightarrow dj}$) sum of the outputs (V_{dj}) of all other neurons, plus an input bias term (I_{di}), i.e.

$$U_{di} = \sum_j T_{di \rightarrow dj} V_{dj} + I_{di} \qquad (1)$$

where $T_{di \to di} = 0$ and $T_{di \to dj} = T_{dj \to di}$. In the original formulation (Hopfield, 1982), all neurons were discrete, i.e., the discrete transfer function g_d, defined below, was used.

$$V_{di} = g_d(U_{di}) = 1 \text{ if } U_{di} > 0$$
$$V_{di} = g_d(U_{di}) = 0 \text{ if } U_{di} < 0 \qquad (2)$$

Note that the notation used in this chapter is slightly different to that used in the original work. The additional subscript d is used to denote variables associated with a discrete neuron, i.e., a neuron using the discrete transfer function (2) above. The following energy function was defined for this system

$$E = -\tfrac{1}{2} \sum_i \sum_j T_{di \to dj} V_{di} V_{dj} - \sum_i I_{di} V_{di} \qquad (3)$$

Thus for a state change in the ith neuron the change in energy is

$$\Delta E = -\left[\sum_j T_{di \to dj} V_{dj} + I_{di}\right] \Delta V_{di} \qquad (4)$$

As the term in brackets is the input to the ith neuron the change in the energy function cannot be positive. Hence, the action of the system is to reduce the energy function and state changes continue to occur until the network has converged to a minimum of this function (3).

6.2.2 CONTINUOUS HOPFIELD NETWORK

The system above was later modified to consider continuous variables [11], i.e., continuous neurons were used. These neurons used the Sigmoid transfer function (5) defined below.

$$V_{ci} = g_c(U_{ci}) = \tfrac{1}{2}(1 + \tanh(\lambda U_{ci})) \qquad (5)$$

In this function λ is a scaling factor termed the slope. The subscript c has been used in place of the subscript d in all variables above. This denotes that the variable is associated with a continuous neuron, i.e., one with a continuous transfer function such as (5). The dynamics of this system were defined by

$$C_i \left[\frac{dU_{ci}}{dt}\right] = \sum_j T_{ci \to cj} V_{cj} - U_{ci}/R_i + I_{ci} \qquad (6)$$

Here C_i and R_i are positive constants based on circuit parameters. It was shown [11] that this system minimises the energy function

$$E = -\tfrac{1}{2} \sum_i \sum_j T_{ci \to cj} V_{ci} V_{cj} - \sum_i I_{ci} V_{ci} + \tfrac{1}{\lambda} \sum_i \left[\left(1/R_i\right) \int_0^{V_{ci}} g_c^{-1}(v) dv \right] \qquad (7)$$

6.2.3 HOPFIELD NETWORKS FOR OPTIMISATION

The Hopfield network has been applied to a wide range of optimisation problems including the travelling salesman problem [12], satellite broadcast scheduling [4] image processing [31] and power system maintenance / planning [9], The usual approach in these methods is to use a penalty function method to define a cost function consisting of the objective function and the equality constraints. This cost function is then mapped to the energy function of the Hopfield network. The network locates a minimum of this energy function as shown above and this will correspond to a minimum of the cost function. To consider inequality constraints inequality constraint neurons are used.

6.2.3.1 Inequality constraint neurons

Tank and Hopfield [26] incorporated inequality constraint neurons into the network. These neurons were used in problems containing linear inequality constraints. However, the neurons were defined incorrectly. The correct formulation was developed by Kennedy and Chua [14].

Inequality constraint neurons operate in a completely distinct fashion to that of other neurons in the network. Consider the linear inequality constraint

$$g(x) \le 0 \qquad (8)$$

where x is the vector of output states of variable neurons. The constraint neuron takes as its inputs the outputs of all the variable neurons, x. The constraint neuron then evaluates the constraint $g(x)$. If the constraint is not violated the constraint neuron produces no output, otherwise an output proportional to the amount of constraint violation is produced. This output is fed back to the inputs of the variable neurons in such a way as to oppose the violation.
Mathematically, if

$$g(x) = b.x \qquad (9)$$

where b is the constraint vector. The constraint neuron output y is given by

$$\begin{aligned} y &= 0 & \text{if } b.x \le 0 \\ y &= K.b.x & \text{if } b.x > 0 \end{aligned} \qquad (10)$$

where K is a positive constant. This constraint neuron output is fed back to the inputs of the variable neurons via the vector $-b$. Hence the input from the constraint neuron to neuron i is given by $-b_i y$, where b_i is the ith element of the vector b. In this fashion constraint neurons encourage variable neurons, whose output contributes to an inequality constraint violation, to adjust their output to alleviate this violation. The amount of encouragement is proportional to the amount of constraint violation and the influence of the particular variable on the constraint. This method has been successfully applied to many problems.

6.2.3.2 Limitations of Hopfield networks

From the above discussion it should be clear that while a Hopfield network may be used to solve an optimisation problem, there are restrictions on the type of problem that may be solved. If the cost function of an optimisation problem is to be mapped exactly to the energy function of a Hopfield network, it must be of the same form as the cost function. This restricts the range of problems to which the network may be applied. It is apparent, for example, that terms of higher order than quadratic may not be mapped to the above functions (3) and (7). It is also clear that two distinct networks have been developed, one containing only discrete terms and one containing only continuous terms. This means that any cost function which contains both discrete and continuous terms may not be accurately mapped to the energy function of a Hopfield network. This is the foremost difficulty in applying Hopfield networks to the scheduling problem.

6.3 Hopfield Networks Applied to the Scheduling Problem

The application of the Hopfield Network to the scheduling problem has many similarities to its application in other problems. The scheduling problem must first be defined in terms of a cost function. This cost function is then mapped to the energy function of a Hopfield Network. The network is initialised and iterated upon until it converges at a minimum of its energy function. There have been many attempts to achieve this and some of these will be described here.

6.3.1 ECONOMIC DISPATCH

In economic dispatch it is required that the optimum operating level of on-line generators be found. The problem is concerned with optimising over continuous variables and hence the continuous version of the Hopfield Network may be used. It will be noticed, however, that the third term in the energy function of continuous Hopfield networks, (7), is not compatible with the cost function of a quadratic programming problem. To eliminate this term from the energy function Park *et al.* [22] redefined the dynamics of the network to be

$$\frac{dU_{ci}}{dt} = \sum_j T_{ci \to cj} V_{cj} + I_{ci} \tag{11}$$

This allowed the energy function for the continuous model to be redefined as

$$E = -\tfrac{1}{2} \sum_i \sum_j T_{ci \to cj} V_{ci} V_{cj} - \sum_i I_{ci} V_{ci} \tag{12}$$

It was shown that the action of the system is to minimise this function by considering its time derivative

A penalty function method was used to represent the economic dispatch problem as a quadratic cost function. Hence to minimise the cost function a cheap solution that produces little or no violation in equality constraints must be found. For the economic dispatch problem the cost function was defined as

$$E = \frac{A}{2}\sum_i \left[a_i P_i^2 + b_i P_i + c_i\right] + \frac{B}{2}\left[L - \sum_i P_i\right]^2 \quad (13)$$

where L is the system demand including transmission losses, P_i is the power output of the ith generator, a_i, b_i and c_i are the cost coefficients of the ith generator and A and B are weighting coefficients. Here the first term in the cost function is the objective function of the optimisation problem and the second term is the load balance equality constraint squared.

It is asserted that by finding a minimum of the cost function, (13), a solution to the economic dispatch problem will be obtained. Maa et al. (1990) illustrated that the equilibrium of such a quadratic programming network is asymptotically stable and is in the neighbourhood of the minimum of the objective function. It was shown that by selecting a sufficiently large penalty factor, B in (13), these points can be made arbitrarily close to each other.

It is necessary to map this cost function (13) to the energy function (12) of the network, i.e., the connection weights of the neural network must be found by setting (13) = (12) This yields the following connection weights.

$$T_{ci\_ci} = -B - Aa_i \quad (14)$$
$$T_{ci\_cj} = -B \quad (15)$$
$$I_{ci} = B.L - Ab_i/2 \quad (16)$$

It is necessary to impose constraints on the variables P_i to prevent them exceeding the capacity limits of the generators. This is achieved by using the modified Sigmoid function

$$V_{ci} = g_{ci}(U_{ci}) = (P_{i,max} - P_{i,min})\left(\frac{1}{2}\left[1 + \tanh(\lambda U_{ci})\right]\right) + P_{i,min} \quad (17)$$

instead of (5).

To implement this algorithm, the neuron inputs, U_{ci}, must be calculated by numerical integration, as (11) only defines the dynamics of the inputs. [22] used a first order Euler-Cauchy integration technique [23] hence the neuron inputs at iteration k were given by

$$U_{ci}(k) = U_{ci}(k-1) + \frac{dU_{ci}}{dt} \quad (18)$$

It was found that convergence of the algorithm typically occurred after approximately 20 000 iterations. [22] also found that, for a given set of parameters, the final solution to the problem did not depend on the initial random neuron outputs since the problem is convex. It was observed that selection of the parameters A and B is not easy and trial and error was used to select them in this work. It was found that there was a slight mismatch between power generated and demand in the solutions found. However, the costs of the schedules compared favourably with those found by a conventional numerical method [33].

[15] applied the Hopfield network to the environmental dispatching problem. Here the optimum operating point of on-line generators such that costs and emissions are minimised is required. Guidelines for selecting some of the weighting parameters were also determined. They illustrated that, given a maximum allowable violation of equality constraints, bounds on the parameter values may be found. The problem of long convergence times was tackled by adding a momentum term to the neuron inputs. This term encourages the neuron inputs to continue moving in the direction they moved at the previous update. Hence the neuron inputs at iteration $k+1$ are given by

$$U_{ci}(k+1) = U_{ci}(k) + \Delta U_{ci}(k) + \eta \Delta U_{ci}(k-1) \tag{19}$$

where the notation $\Delta U_{ci}(k)$ has replaced the notation $\dfrac{dU_{ci}}{dt}$ and η is a factor experimentally selected to be 0.95. Results showed that the method proposed here failed to find solutions as cheap as those found with the Newton Raphson method. It was also found that including transmission losses added to the difficulty of the problem.

[3] applied a Hopfield network to the advance dispatch problem using a transputer network. The dynamics described above are used in this implementation. Gaussian noise is added to all inputs to avoid convergence to a local minimum. In this work, units are dispatched for 13 time intervals in the 0-60 minutes ahead range using a load forecast for every five minutes. When this is complete the dispatch for five minutes ahead is used and the process repeated to find a dispatch for the next time step. The objective function, below, is slightly different to those above as it relates to more than one time period

$$\sum_{j=2}^{m}\sum_{i} a_i P_{ij}^2 + b_i P_{ij} + c_i \tag{20}$$

where P_{ij} is the power output of unit i at time j. The summation over time begins at $j = 2$ since $j = 1$ represents the present time (0 minutes ahead). This application considers the power balance, upper and lower generator limits, ramp rate and operating point constraints. The operating point constraints ensure that generator outputs are at levels such that future loads can be met. As in previous work, generator limits are solved by the modified Sigmoid function (17) and the load balance constraint is incorporated into the cost function.

The other constraints are taken together and a feasible region is defined for each generator at each time interval m, i.e., if the operating point is outside the feasible region, these constraints could not be satisfied in either interval $m-1$ or interval $m+1$. It was noted [3] that ôthe solution can be converged gradually to a feasible suboptimum result" using this method. To help avoid convergence to a local minimum the initial values of the neurons were set to the results obtained in the previous run shifted backwards by one time interval. The results obtained by this method were very similar to those obtained from quadratic programming.

[16] included loss penalty factors to represent transmission losses in the system. The penalty factors were calculated by a state estimation program. A clamped state variable technique was used to reduce the convergence time of the Hopfield network. Results showed that this method found results close to those achieved by numerical methods.

[25] proposed a direct computation method for solving the economic dispatch problem of thermal generators with Hopfield networks. By examining the differential equations that govern the dynamics of the system, the final values of problem variables were determined. This greatly reduces the computation time required by the algorithm. However, the solution equations so obtained resemble those used in Newton-Raphson algorithms. The authors also present a method to assist selection of penalty parameters similar to that described in [15].

6.3.2 UNIT COMMITMENT

In unit commitment, the on/off schedule of generators over a time period must be determined. In [24] the discrete version of the Hopfield network was applied to this problem. The neurons were arranged in a two dimensional grid where the output of neuron *dij*, V_{dij}, represents the status of generator i at time j, i.e.,

$$V_{dij} = 1 \Rightarrow \text{generator is on-line}$$
$$V_{dij} = 0 \Rightarrow \text{generator is off-line} \quad (21)$$

The cost function was defined as

$$\sum_i \sum_j \left[V_{dij} C_i(P_{ij}) + S_i V_{dij}(1 - V_{di,j-1}) \right] \quad (22)$$

where S_i is the startup cost of generator i and $C_i(P_{ij})$ is the fuel cost associated with the power generated by generator i at time j.

Since discrete neurons were used in this implementation, the variables P_{ij} were not specified. This means that the fuel cost above could not be accurately computed and the load balance constraint could not be evaluated accurately. To overcome these problems it was assumed that the fuel cost of any on-line generator was equal to the fuel cost it would have if it were operating half way between its maximum and minimum capacity limits and constraint neurons were used to implement load balance. These approximations severely distorted the problem the network was attempting to solve.

Hence the objective function was

$$\sum_i \sum_j \left(S_i V_{dij}(1 - V_{dij-1}) \right) + f_i V_{dij} \qquad (23)$$

where f_i was taken to be the fuel cost of the ith unit if it was on line i.e.

$$C_i (P_{ij}) = f_i \qquad (24)$$

Note that this assumes that the fuel cost of a unit is independent of its power output. Since the power output of on-line generators is not specified it is not possible to enforce the load balance constraint with a penalty function method as above. Instead [24] posed the load balance and spinning reserve constraints as a single inequality constraint of the form

$$\sum_i P_{i,max} V_{dij} - L_j - R_j \geq 0 \qquad \forall j \qquad (25)$$

where R_j is the reserve required at hour j. This constraint ensures sufficient capacity is on-line at all times to meet the load and reserve. It does not specify how the load is distributed among on-line units. A similar constraint to ensure that the sum of minimum capacities of on-line units is less than system demand was also included. These two constraints were enforced using inequality constraint neurons. Minimum up and down times were also considered using inequality constraint neurons.

[24] applied this method to a unit commitment problem with 30 units and 24 time periods. The network was randomly initialised 100 times and results were distributed in the range 265 - 271 (cost, units not supplied). This illustrates that the network has many local minima and that multiple runs from different initial conditions may improve solution quality. Lagrangian relaxation was applied to the same problem and a result of 263.35 was obtained. This is 0.6 % less expensive than the best solution found by the Hopfield network. [24] suggested that the major cause for more expensive solutions may be the inexact nature of the mapping of the unit commitment problem to the neural network. It was further suggested that a more exact mapping would lead to a more complex neural network. Achieving convergence in such a network was identified as, "truly a big challenge for future studies".

6.3.3 COUPLED GRADIENT NETWORKS

Watta and Hassoun defined a coupled gradient network (CGN) [32] and applied it to mixed integer programming. The CGN attempted to overcome the most serious limitation of previous neural network attempts to solve such problems, in that all problem variables were now specified. In a CGN there are two sets of neurons, one representing discrete variables, a v-network, and one representing continuous variables, an x-network. The input to the ith neuron in the x-network is denoted h_i^x and the input to the ith neuron in the v-network, h_i^v. Both sets of neurons use the

Sigmoid non-linearity as their transfer function. Consider the generalised mixed integer programming problem

$$\text{Min} \quad f(x,v)$$
$$\text{subject to } g_i(x,v) \leq 0 \quad \forall i = 1.....m$$
$$h_i(x,v) = 0 \quad \forall i = 1.....p$$
$$v_i \in \{0,1\} \quad \forall i = 1.....q \quad (26)$$

where there are m inequality constraints, p equality constraints and q integer variables. The energy function for this problem was defined as

$$E(x,v) = f(x,v) + P(x,v) \quad (27)$$

where $f(x,v)$ is the objective function and $P(x,v)$ is a penalty function term comprising of three terms

$$P(x,v) = G(x,v) + H(x,v) + V(v) \quad (28)$$

Here $G(x,v)$ imposes a penalty for any violated inequality constraints, $H(x,v)$ penalises violations of equality constraints and $V(v)$ is a combinatorial constraint included to encourage the v variables to saturate at either 1 or 0. This final term is necessary as continuous values are being used to represent discrete variables. $G(x,v)$ takes the form

$$G(x,v) = \sum_{i=1}^{m} \varphi[g_i(x,v)] \quad (29)$$

where $\varphi(\varepsilon) = \varepsilon^2$ if $\varepsilon > 0$ and $\varphi(\varepsilon) = 0$ for all $\varepsilon \leq 0$. This term acts almost identically to the inequality constraint neurons described earlier. $H(x,v)$ is defined as

$$H(x,v) = \sum_{j=1}^{p} h_j^2(x,v) \quad (30)$$

and finally

$$V(v) = \sum_{i=1}^{q} v_i(1 - v_i) \quad (31)$$

The dynamics of the CGN are defined as

$$\dot{h}_i^x = -\eta_x \left[A \frac{\partial f}{\partial x_i} + B \sum_{j=1}^{m} \varphi'[g_j(x,v)] \frac{\partial g_j}{\partial x_i} + C \sum_{j=1}^{p} 2h_j(x,v) \frac{\partial h_j}{\partial x_i} \right] \quad (32)$$

$$\dot{h}_i^v = -\eta_v \left[A \frac{\partial f}{\partial v_i} + B \sum_{j=1}^{m} \varphi'[g_j(x,v)] \frac{\partial g_j}{\partial v_i} + C \sum_{j=1}^{p} 2h_j(x,v) \frac{\partial h_j}{\partial v_i} + D(1 - 2v_i) \right]$$

$$(33)$$

where η_x and η_v are positive coefficients used to scale the dynamics of the network. The neuron transfer functions are

$$x_i = \rho_i(h_i^x) \quad (34)$$

$$v_i = \sigma_i\left(h_i^y\right) \tag{35}$$

here A, B, C and D are positive penalty coefficients as in previous cost function formulations. Although all transfer functions, ρ_i and σ_i, are Sigmoidal, separate symbols are used as different slopes will be used for discrete and continuous variables. It can be shown [36] that the dynamics so defined cause the CGN to seek out a minimum of the energy function (27).

The CGN as described above was applied to the unit commitment problem. In this implementation the status of each unit was represented by a v neuron while the power output was represented by an x neuron. The problem considered the load balance and spinning reserve constraints. The parameters A, B, C, D and the slopes of the Sigmoid function were established by trial and error, as it was noted that there is no theoretically established method for finding them.

Table 1 shows the results obtained when the CGN was compared directly with Lagrangian Relaxation. These results show that the best cost obtained by the CGN over 20 runs on a ten unit thermal system with time horizons, t, from 5 to 24 hours. The figure in brackets is the % difference between the two costs. Also shown is the percentage of CGN solutions which are feasible.

TABLE 1 Comparison of CGN with LR

	t=5	t=10	t=15	t=20	t=24
Lagrangian Relaxation	11 087	21 824	31 345	39 797	47 511
CGN	11 147 (0.5)	22 076 (1.2)	31 912 (1.8)	40 244 (1.1)	48 498 (2.1)
Percentage Feasible solutions CGN	100	100	80	70	55

It can be seen that even over multiple runs the CGN is unable to provide a better solution than Lagrangian Relaxation. However, of more concern is the increasing cost relative to LR as the problem size increases. It can also be seen that as the time horizon increases the percentage of feasible solutions obtained decreased, for example, as the scheduling horizon increased from 15 hours to 24 hours the percentage of feasible solutions obtained decreased from 80% to 55% for a 10 generator problem. It was suggested [36] that, "convergence to interior points of the hypercube is one of the reasons why the coupled net may produce solutions which are not feasible". It was found that in many cases the v neurons converged to a point between 0 and 1 and hence must be thresholded to determine the unit's status. This illustrates two inadequacies in the approach of the CGN. Firstly discrete costs and constraints, e.g., start-up costs may not be accurately modelled by a continuous variable and secondly it is difficult to achieve a feasible schedule when there is this level of uncertainty in the solution provided.

In a real power system, as more units and longer planning horizons are commonly studied, the trends of less feasible solutions and diminishing quality of solution with increasing problem size illustrate that the CGN may not be suitable for real problems.

In [20] the unit commitment problem was also tackled using a coupled network. However, a Boltzmann machine was used for the discrete variables. This means a stochastic update rule was used. This does not guarantee a decreasing energy function but may avoid local minima. Results were given for scheduling 20 thermal units but as no comparison was made with other methods it is difficult to evaluate solution quality.

6.4 Augmented Hopfield Network

To accurately model the scheduling problem, an algorithm must consider both discrete and continuous variables. None of the Hopfield style algorithms described above achieve this. The difficulty in including both types of variables in one approach is significant. State transition arguments used in discrete Hopfield networks can not be applied to continuous variables, and dynamic arguments used in continuous Hopfield networks can not be applied to discrete variables as they are not differentiable.

It is desirable that all problem constraints be considered simultaneously, as "unless a global optimisation technique is used the final solution will be sub-optimal" [30]. The Augmented Hopfield Network [27][28] is an algorithm that contains all these features and successfully overcomes the difficulties encountered in previous attempts to apply the Hopfield network to the scheduling problem.

6.4.2 NETWORK ARCHITECTURE

In the AHN there are two sets of neurons, a discrete set and a continuous set. These neurons are arranged in pairs, with every discrete neuron having an associated continuous neuron and vice versa. The output of discrete neuron ij, is V_{dij}, and the output of continuous neuron ij, is V_{cij}. All neurons have an input bias, I_{dij} for discrete neuron ij and I_{cij} for continuous neuron ij. The standard interconnection matrix T is used, giving connections between all neurons, discrete and continuous. For example, T_{dij_ckm} is the connection from the discrete neuron ij to the continuous neuron km. A new form of interconnection (a matrix W) between neuron pairs is introduced. For example W_{ij_km} is the connection from neuron pair (dij, cij) to neuron pair (dkm, ckm). Figure 1 illustrates the influence of this new connection weight on the continuous neuron km. This diagram is merely a schematic and the effects of network dynamics are not shown.

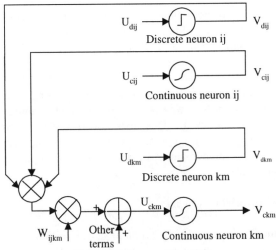

Figure 1 New connection weight

The output of each neuron is given by
$$V_{dij} = g_d(U_{dij}) \tag{36}$$
$$V_{cij} = g_c(U_{cij}) \tag{37}$$
where U_{dij} and U_{cij} are the inputs of discrete neuron ij and continuous neuron ij respectively. Discrete neurons use the discrete transfer function, g_d, given by

$$V_{dij} = g_d(U_{dij}) = 1 \quad \text{if } U_{dij} > 0$$
$$V_{dij} = g_d(U_{dij}) = 0 \quad \text{if } U_{dij} < 0$$
$$\text{No change in } V_{dij} \quad \text{if } U_{dij} = 0 \tag{38}$$

Continuous neurons use the Sigmoid transfer function, g_c,

$$V_{cij} = g_c(U_{cij}) = \tfrac{1}{2}\left(1 + \tanh(\lambda U_{cij})\right) \tag{39}$$

where λ is a scaling factor known as the slope.

6.4.3 NETWORK DYNAMICS

The dynamics of the augmented model are defined by equations (40) and (41) below.

$$\frac{dU_{cij}}{dt} = \sum_{k,m} T_{ckm \to cij} V_{ckm} + \sum_{k,m} T_{dkm \to cij} V_{dkm} + I_{cij} \\ + \sum_{k,m} W_{km \to ij} V_{dij} V_{dkm} V_{ckm} \tag{40}$$

$$U_{dij} = \begin{pmatrix} \sum_{k,m} T_{ckm \to dij} V_{ckm} + \sum_{k,m} T_{dkm \to dij} V_{dkm} + I_{dij} + \sum_{k,m} W_{km \to ij} V_{ckm} V_{dkm} V_{cij} \\ + \frac{1}{2} T_{dij \to dij} \Psi + \frac{1}{2} W_{ij \to ij} V_{cij}^2 \Psi \end{pmatrix}$$

(41)

where
$\Psi = -1$ if $V_{dij} = 1$
$\Psi = 1$ if $V_{dij} = 0$ (42)

The energy function, (43), below is proposed.

$$E = -\frac{1}{2} \sum_{i,j} \sum_{k,m} T_{ckm \to cij} V_{cij} V_{ckm} - \sum_{i,j} \sum_{k,m} T_{dkm \to cij} V_{cij} V_{dkm} - \frac{1}{2} \sum_{i,j} \sum_{k,m} T_{dkm \to dij} V_{dij} V_{dkm}$$
$$- \sum_{i,j} I_{cij} V_{cij} - \sum_{i,j} I_{dij} V_{dij} - \frac{1}{2} \sum_{i,j} \sum_{k,m} W_{km \to ij} V_{dij} V_{cij} V_{dkm} V_{ckm}$$

(43)

It can be shown [28] that the dynamics of the system described above cause this energy function (43) to decrease.

The dynamics of the system were defined by (40) and (41). To calculate the continuous neuron inputs, (U_{dij} and U_{cij}), these equations must be numerically integrated. Using first order Euler-Cauchy integration [23] with a step size of one, the input to each continuous neuron at iteration n is

$$U_{cij}(n) = U_{cij}(n-1) + \sum_{k,m} T_{ckm \to cij} V_{ckm} + \sum_{k,m} T_{dkm \to cij} V_{dkm} + I_{cij}$$
$$+ \sum_{k,m} W_{km \to ij} V_{dij} V_{dkm} V_{ckm}$$

(44)

The AHN is said to have converged when no discrete neuron changes its state and no continuous neuron changes its output by more than a specified tolerance, on successive iterations. An iteration is said to have occurred when all neurons in the AHN have updated their states.

As the AHN minimises a function consisting of both discrete and continuous terms within a single algorithm, the AHN may be used to solve any general mixed integer optimisation problem with linear and quadratic terms. This is a very general class of problem that occurs often in real world applications.

6.5 Application of the Augmented Hopfield Network to the Scheduling Problem

To apply the AHN to the scheduling problem, the objective function and equality constraints are combined to form a cost function which is mapped to the energy function described above and then minimised by the network. Inequality constraints are meanwhile tackled by dedicated constraint neurons and other approaches described in this section.

6.5.1 COST FUNCTION

In this section a complete cost function for the scheduling problem of a power system with thermal, hydro and pumped storage units is developed as a penalty function solution method is used by the AHN This cost function consists of the objective function (fuel costs, idling costs and start-up costs) and equality constraints. The various terms in the cost function will be scaled by weighting coefficients (*A*,*B*,*C*,*D*) as in previous work [22]. The cost function is written in terms of the AHN variables V_{dij} and V_{cij} where V_{dij} represents the status of the ith generator at time j (i.e. $V_{dij} = 1$ if the generator is on-line and $V_{dij} = 0$ if the generator is off-line) and where V_{cij} is the power output of the generator if it is on-line. Hence the product $V_{dij}V_{cij}$ always represents the actual power output of the ith generator at time j.

6.5.1.1 Objective function.
The objective function of the scheduling problem is the financial cost of the schedule. It is assumed that the fuel cost of all units is quadratic with coefficients a_i, b_i and c_i and that the start-up cost, S_i, is independent of the time off-line. Hence the first term of the cost function is

$$A\left[\begin{array}{l}\sum_i\sum_j\left(S_i V_{dij}(1-V_{dij-1})\right)\\+\sum_i\sum_j\left(a_i\left(V_{dij}V_{cij}\right)^2+b_i V_{dij}V_{cij}+c_i V_{dij}\right)\end{array}\right] \quad (45)$$

where *A* is a weighting parameter.

6.5.1.2 Hydro constraints.
Here the efficiency of a hydro unit is assumed to be a constant value for the hydro unit's entire operating region. Regions of instability and high inefficiency may be excluded by adjusting the operating limits of the generator. The water constraint may be considered as an equality constraint since it is economical to generate the maximum possible power available from each unit. This leads to a target, Θ_i, being defined for each hydro unit and deviations from this target are penalised by the following term in the cost function

$$C\sum_{i\in H}\left(\sum_j V_{dij}V_{cij}-\Theta_i\right)^2 \quad (46)$$

where C is a weighting parameter as above and H denotes the set of all hydro units. It is advantageous to consider hydro units in this fashion since an element of dispatch may be performed if they are dealt with in the commitment time frame. Some conventional formulations assume that the total amount of time that units may be committed for is the total amount of time that they can be run at maximum output. In the above formulation, the actual power output at each hour is specified by the scheduling algorithm and so a more flexible approach to the total run time of a unit exists and more optimal schedules may be found.

6.5.1.3 Pumped storage units.

For pumped storage stations it is an operational requirement that all water used be replaced within a given study period. Thus while there is no direct cost associated with generation from these units, there is an indirect cost associated with replacing water used. These factors are considered by modelling each unit as two individual units, a pump and a generator. For a generation/pumping efficiency of α_i the following term is added to the cost function.

$$D \sum_h \left(\sum_{i \in PG_h} \sum_j V_{dij} V_{cij} - \sum_{i \in PP_h} \sum_j \alpha_i V_{dij} V_{cij} \right)^2 \quad (47)$$

where PG_h is the set of pumped storage units operating as generators and PP_h is the set of pumped storage units operating as pumps at reservoir h and D is a weighting parameter. Note that as operating modes are modelled as separate units, each mode of each generator has its own unique index i.

For completeness it is also necessary to include a constraint to prevent any of the modes operating simultaneously. To achieve this, the inputs to the discrete neurons representing all operating modes of a unit at a given time are evaluated. If more than one of these are positive, a reduction in the energy function would occur if all the modes with positive inputs to discrete neurons were allowed to come on-line. However, as this is not permitted in practice, only the mode which gives the greatest reduction in energy is allowed to come on-line. It can be shown that the greatest reduction in energy comes from bringing the mode with the largest discrete input on-line (Walsh et al.,1998). -This ability to model different operating modes of a generator is a particularly useful feature of the AHN applied to generator scheduling.

6.5.1.4 Load balance.

The load balance constraint must take account of the fact that generation must be provided for units operating as pumps as well as the load and losses, L_j, at time j. The penalty term below is hence added to the cost function of the scheduling problem.

$$B\sum_j \left(\sum_{i \notin PP} V_{dij} V_{cij} - \sum_{i \in PP} V_{dij} V_{cij} - L_j\right)^2 \qquad (48)$$

where PP is the set of all units operating in pumping mode in the system and B is a weighting parameter.

6.5.1.5 Fuel constraints.

In binding fuel constraints it is possible to define a target amount of energy to be produced by a unit or group of units over a given time frame. This can be treated as an equality constraint and incorporated into the cost function. For a single unit i, constrained to generate Θ_i MWhrs of energy over the study period the following term (49) is added to the cost function.

$$F\left(\sum_j V_{dij} V_{cij} - \Theta_i\right)^2 \qquad (49)$$

where F is a weighting coefficient. In practice this implementation is similar to modelling the unit as a hydro unit with a start up and fuel cost. This is an accurate representation of a binding fuel constraint. If there are multiple fuel constraints and it is not possible to determine in advance which ones will be binding, dedicated constraint neurons may be used to implement this constraint.

6.5.1.6 Total cost function.

The total cost function described above may be written as

$$\begin{aligned}
\text{Cost} = &\ A\left[\sum_i \sum_j \left(S_i V_{dij}(1 - V_{dij-1})\right) + \sum_i \sum_j \left(a_i \left(V_{dij} V_{cij}\right)^2 + b_i V_{dij} V_{cij} + c_i V_{dij}\right)\right] \\
&+ B\sum_j \left(\sum_{i \notin PP} V_{dij} V_{cij} - \sum_{i \in PP} V_{dij} V_{cij} - L_j\right)^2 + C\sum_{i \in H}\left(\sum_j V_{dij} V_{cij} - \Theta_i\right)^2 \\
&+ D\sum_h \left(\sum_{i \in PG_h} \sum_j V_{dij} V_{cij} - \sum_{i \in PP_h} \sum_j \alpha_i V_{dij} V_{cij}\right)^2 + F\left(\sum_j V_{dij} V_{cij} - \Theta_i\right)^2
\end{aligned}$$
(50)

It is this function that must be minimised to find a solution to the scheduling problem.

6.5.2 MAPPING COST FUNCTION TO ENERGY FUNCTION

The cost function developed above includes the objective function and the equality constraints of the scheduling problem. To minimise this using the AHN it is first necessary to map the cost function to the energy function. The energy function

(44) contains terms of the same form as every combination of discrete and continuous variable found in this cost function. Thus the cost function, can be mapped to this energy function i.e. $E = Cost$. This exact mapping was not possible with other versions of the Hopfield network as their energy functions did not include both discrete and continuous terms

Hence by locating a minimum of the energy function, the network will have found a minimum of the cost function. Where there is an intersection in the connection weights, found above, the actual weight is the sum of the individual weights. The full set of connection weights and input biases are shown below. All connection weights are symmetric, i.e., $T_{dij \rightarrow dkm} = T_{dkm \rightarrow dij}$, $T_{dij \rightarrow ckm} = T_{ckm \rightarrow dij}$, $T_{cij \rightarrow ckm} = T_{ckm \rightarrow cij}$ and $W_{ij \rightarrow km} = W_{km \rightarrow ij}$. These indices take on all possible values unless otherwise stated. All weights and biases not defined below are zero.

$$T_{dij \rightarrow cij} = -Ab_i + 2BL_j + 2F\Theta_i \quad i \notin H, i \notin PP, i \in FC \quad (51)$$
$$T_{dij \rightarrow cij} = -Ab_i + 2BL_j \quad i \notin H, i \notin PP, i \notin FC \quad (52)$$
$$T_{dij \rightarrow cij} = -Ab_i + 2C\Theta_i + 2BL \quad j \in H \quad (53)$$
$$T_{dij \rightarrow cij} = -Ab_i - 2BL_j \quad i \in PP \quad (54)$$
$$I_{dij} = -AS_i - Ac_i \quad (55)$$
$$T_{dij-1 \rightarrow dij} = AS_i \quad (56)$$
$$W_{ij \rightarrow ij} = -2Aa_i - 2B \quad i \notin H \quad (57)$$
$$W_{ij \rightarrow ij} = -2Aa_i - 2C - 2B \quad i \in H \quad (58)$$
$$W_{ij \rightarrow im} = -2C \quad i \in H, m \neq j \quad (59)$$
$$W_{ij \rightarrow im} = -2F \quad i \in FC, m \neq j \quad (60)$$
$$W_{ij \rightarrow kj} = -2B \quad i,k \notin PP, i \neq k \quad (61)$$
$$W_{ij \rightarrow kj} = -2B \quad i,k \in PP \quad (62)$$
$$W_{ij \rightarrow kj} = 2B \quad i \in PP, k \notin PP \quad (63)$$
$$W_{ij \rightarrow km} = -2D \quad i,k \in PG_h \quad (64)$$
$$W_{ij \rightarrow km} = 2\alpha_i D \quad i \in PG_h, k \in PP_h \quad (65)$$
$$W_{ij \rightarrow km} = -2\alpha_i^2 D \quad i,k \in PP_h \quad (66)$$

As a penalty function method is used, equality constraints may not be satisfied exactly. It was observed that for a given set of parameters the amount of violation of equality constraints is a constant. This result is intuitive. Ideally the minima of the cost function would occur at points where the equality constraints are completely satisfied. In practice, violation of a constraint results in an increase in the relevant penalty term of the cost function but may cause a reduction in the objective function (45). For example under-serving the load will cause an almost linear reduction in fuel costs for most systems as well as a quadratic increase in the penalty term (48). These two effects cancel almost exactly at a certain level of constraint violation. This small offset level is easily determined experimentally for a given system and set of parameter values and may be factored in to load forecasts, hydro targets etc. to cancel the effect. This practice may prove useful in a wide range of applications where penalty function methods are used.

6.5.3 INEQUALITY CONSTRAINTS

It has been illustrated above that the AHN may be applied to the scheduling problem with equality constraints. However, there are a range of inequality constraints which must be observed in practice by scheduling algorithms. A number of different approaches will be taken to solve these constraints.

6.5.3.1 Spinning reserve
The spinning reserve constraint can be tackled by dedicated constraint neurons [14]. These neurons have the following input,

$$U_{bj} = \sum_i r_i(P_{ij}) - R_j \qquad (67)$$

where $r_i(P_{ij})$ is the reserve contribution of generator i as a function of its current power output, the subscript b is used to denote a constraint neuron and R_j is the reserve required at hour j. The output of the constraint neurons is given by

$$V_{bdij} = V_{bcij} = 0 \quad \text{if} \quad U_{bj} \geq 0$$
$$V_{bdij} = -K_{di}(U_{bj}) \quad \text{if} \quad U_{bj} < 0 \qquad (68)$$

where K_{di} is a constant for each discrete variable neuron to which the constraint neuron is connected. If the constant K_{di} is positive for a variable neuron the input to that variable neuron is increased when the constraint is violated. Conversely if the constant is negative for a variable neuron the input to that variable neuron is reduced when the constraint is violated. This property is used to select K_{di} such that the outputs of constraint neurons are applied to the inputs of variable neurons to encourage them to provide more reserve when required. If the units have the "straight back" reserve characteristic, i.e.,

$$r_i(P_{ij}) = P_{i,max} - P_{ij} \qquad (69)$$

then the constant K_{di} is positive. This means that if the reserve constraint is violated all generators are encouraged to come on-line. Note that there is no benefit in reducing power output of a given generator as this power must be replaced by other generators and hence the total reserve remains unchanged. The actual distribution of the reserve amongst the units will be decided by the dynamics of the system.

6.5.3.2 Generator capacity limits.
The maximum ($P_{i,max}$) and minimum ($P_{i,min}$) generation limits of the ith generator are tackled by using the modified Sigmoid function (70) as defined in (Park et al., 1993), for all continuous neuron transfer functions.

$$V_{cij} = \left[(P_{i,max} - P_{i,min}) * \tfrac{1}{2}(1 + \tanh(\lambda U_{cij}))\right] + P_{i,min} \qquad (70)$$

As this function is monotonic increasing the convergence proof remains valid.

6.5.3.3 Minimum up and down times.

In [24] minimum up and down times (*mut, mdt*) were considered by inequality constraint neurons. This approach is biased as not all violated constraints are penalised equally. In the AHN discrete neurons may be iterated upon in a sequential (non random) fashion. When a generator start-up is detected at hour *j* the generator is not permitted to come on-line unless keeping it on-line for hours *j, j+1,...., j+mut-1* would result in a net reduction of the energy function. If this is the case, the generator is brought on-line for these hours and iterations continue at hour *j+mut*, otherwise the generator remains off-line at hour *j* and iterations continue at hour *j+1*. This will occur every time these neurons are iterated upon thus ensuring that all generators obey minimum up time constraints. A similar approach was taken for minimum down times.

The above method is not completely satisfactory as certain optimal combinations may not be detected. This occurs as the algorithm only considers the case of bringing the generator on-line for its minimum up time and not for longer periods. There may exist cases when it may not be beneficial to turn a generator on for its minimum up time but it may be beneficial to turn it on for a longer period. Consider the example of a thermal generator with a minimum up time of 4 hours. If the following series of inputs exist at discrete neurons representing this generator at adjacent time intervals

-100, 20, 20, 20, -70, 20, 20, 20, -100

the optimal strategy is to bring the generator on-line for the middle seven intervals. However, as the strategy above only searches four adjacent periods this solution is not found and the generator remains off-line. This problem could be overcome by searching more combinations of strategies when evaluating the constraint. However, due to the nature of the scheduling problem single interval "spikes" such as that seen above, (-70), rarely occur and hence the strategy above is adequate in most cases. For this reason the additional computational burden of checking multiple strategies may be avoided.

6.5.3.4 Operating ramp rates.

Ramp rate limits are tackled by dedicated constraint neurons. For operating ramp constraints these neurons connect every continuous neuron to the continuous neurons representing the same generator in adjacent time periods. Before a continuous neuron is updated the constraint neurons linking it to the neurons representing adjacent hours must first be updated. The input to the constraint neuron for operating ramp-up rates from hour *j-1* to hour *j* for generator *i* is given by

$$U_{ruij} = (V_{cij} - (V_{cij-1} + rur_i)) \tag{71}$$

where rur_i is the ramp up rate of unit i and the subscript $ruij$ indicates a ramp up constraint on generator i at time j. This constraint neuron has two distinct outputs, V^{cij}_{ruij} and V^{cij-1}_{ruij}. The output V^{cij}_{ruij} is connected back to the input of the continuous neuron i,j to encourage it to reduce its output if the ramp up constraint is violated. The output V^{cij-1}_{ruij} is simultaneously connected to the input of continuous neuron $i,j-1$ to encourage it to increase its output if the constraint is violated. These outputs are given by

$$V^{cij}_{ruij} = 0 \quad \text{if } U_{ruij} < 0$$
$$= -L.U_{ruij} \quad \text{if } U_{ruij} > 0 \qquad (72)$$

$$V^{cij-1}_{ruij} = 0 \quad \text{if } U_{ruij} < 0$$
$$= L.U_{ruij} \quad \text{if } U_{ruij} > 0 \qquad (73)$$

where L is a positive constant. In this manner both variables involved in constraint violation are equally encouraged to move towards a feasible solution without bias. A graphical illustration of this constraint is provided in figure 2

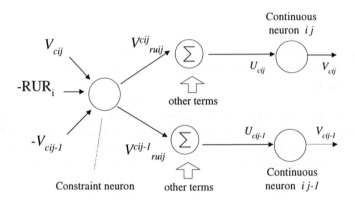

Figure 2 Operating ramp up constraint neuron for hour $j-1, j$

The constraint neurons for ramp down rates operate in a similar fashion. The input to the constraint neuron for operating ramp-down rates from hour $j-1$ to hour j for generator i is given by

$$U_{rdij} = (V_{cij} - (V_{cij-1} - rdr_i)) \qquad (74)$$

where rdr_i is the ramp down rate of unit i and the subscript $rdij$ denotes a ramp down constraint on generator i at time j. The neuron's outputs are

$$V^{cij}_{rdij} = 0 \quad \text{if } U_{rdij} > 0$$
$$= -L.U_{rdij} \quad \text{if } U_{rdij} < 0 \qquad (75)$$

$$V^{cij-1}_{rdij} = 0 \quad \text{if } U_{rdij} > 0$$
$$= L.U_{rdij} \quad \text{if } U_{rdij} < 0 \qquad (76)$$

Here V^{cij}_{rdij} is fed back to the input of continuous neuron i,j causing it to increase its output if the ramp down constraint is violated, while V^{cij-1}_{rdij} is connected to the input of continuous neuron $i,j-1$ encouraging it to decrease its output in the case of a violated constraint.

6.5.3.5 Start-up and shut-down ramp rates.

For start-up and shut-down ramp constraints it is usual to have the generators confined to a set trajectory or a restricted cone of trajectories [17]. When a start-up or shut-down is detected, the relevant continuous neurons must be clamped to values that follow this trajectory until the unit reaches minimum power or is fully off-line. To prevent this method from suffering from temporal [30], it is necessary to include constraint neurons between discrete and continuous neurons at the state transition. This allows information regarding the state transition to propagate temporally throughout the solution algorithm.

For example, consider the case where the algorithm, without ramp constraints, would schedule a generator at a level close to its maximum rated capacity, but operating ramp limits constrain it to a much lower level. This unit is on-line at adjacent hours and the constraint neurons between continuous neurons, as described above, simultaneously encourage the output to increase at the limiting adjacent hour and to decrease at the present hour. However, if the generator is off-line at an adjacent hour and is constrained to operate at a low level by start-up/shut-down ramp constraints, the generator output is clamped to a low value and cannot increase its output unless the relevant adjacent discrete neuron changes its state. Hence constraint neurons are required to act between discrete and continuous neurons in this case. The input to these neurons for a generator startup will be

$$U_{ru\;dij} = (V^u_{cij} - V^c_{cij}) \qquad (77)$$

where V^u_{cij} is the unconstrained output of continuous neuron i,j and V^c_{cij} is the constrained output. The output of this neuron is

$$V^{dij-1}_{ru\;dij} = L \cdot U_{ru\;dij} \qquad (78)$$

where L is a positive constant. In this case a generator that was scheduled to come on-line at hour j is constrained to have a set power output, V^c_{cij}, at this hour. If the dynamics suggest this generator should be at a different level, V^u_{cij}, this information is fed back to the discrete neuron at hour $j-1$, (78), encouraging the generator to come on line earlier thereby permitting the desired higher level of generation to be reached at hour j. A graphical illustration of this constraint is provided in figure 3 below. Similar constraint neurons act at a generator shut-down.

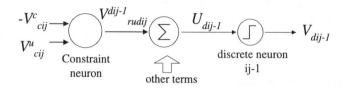

Figure 3 Start-up ramp constraint neuron at hour *j-1*.

6.5.3.6 Transmission constraints.

Dedicated constraint neurons are also used for transmission constraints. The input to the constraint neuron for the upper limit on line m is

$$\overline{U_{mj}} = \overline{P_{mj}} - \sum_i k_{mi} V_{dij} V_{cij} \tag{79}$$

similarly for the lower limit

$$\underline{U_{mj}} = \underline{P_{mj}} - \sum_i k_{mi} V_{dij} V_{cij} \tag{80}$$

The outputs of these neurons are

$$\begin{aligned}
\overline{V_{mij}} &= 0 &&\text{if } \overline{U_{mj}} > 0 \\
&= J.k_{mi}.\overline{U_{mj}} &&\text{if } \overline{U_{mj}} < 0 \quad (81) \\
\underline{V_{mij}} &= 0 &&\text{if } \underline{U_{mj}} < 0 \\
&= J.k_{mi}.\underline{U_{mj}} &&\text{if } \underline{U_{mj}} > 0 \quad (82)
\end{aligned}$$

where J is a positive constant analogous to K. These outputs are connected to the inputs of the continuous neurons representing the ith generator at the jth hour. In this manner if a transmission line capacity constraint is violated, the generators are encouraged to adjust their outputs in a manner that will help to alleviate this constraint violation. It is necessary to also provide these inputs to the corresponding discrete neurons, as the offending generators may be at their maximum or minimum capacity limits and a state change may be necessary to find a feasible solution.

6.5.4 INITIALISATION OF AHN

The AHN is initialised by setting the neuron outputs to random numbers leading to a distribution of results when the algorithm is run multiple times. As a result, better solutions will be obtained if the algorithm is run a large number of times. This leads to an undesirable increase in the effective running time of the algorithm. It has been seen that the LR algorithm for the generator scheduling problem involves solving a related dual problem. The algorithm produces low cost solutions within a very short computation time. However, these results are generally not feasible with respect to relaxed constraints and must be perturbed, frequently by the use of heuristic techniques [8][35], to achieve a feasible solution.

An alternative technique for the initialisation of the AHN [1] is described here. An infeasible Lagrangian dual maximum solution is used to initialise the AHN, i.e., the neuron outputs will be initialised with the infeasible dual maximum LR schedule. This will allow the AHN to take advantage of the information gained from the application of the LR algorithm. From equation (44) continuous neuron inputs at iteration n depend on the inputs at iteration $n-1$. However, the LR schedule only provides the neuron outputs and so the set of neuron inputs that would have yielded this state must be evaluated.

The transfer function for continuous neurons is bijective, (70). Hence given a continuous neuron's output, its unique input may be found by renormalising the output and evaluating the inverse transfer function (83).

$$U_{cij} = \frac{-1}{2\lambda} \ln\left(\frac{1-V_{cij}}{V_{cij}}\right) \tag{83}$$

The inverse is not required for discrete neurons since the input to a discrete neuron is independent of the inputs at previous iterations (41).

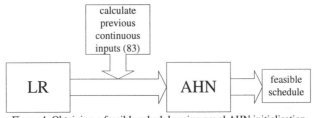

Figure 4 Obtaining a feasible schedule using novel AHN initialisation

The network, now initialised, iterates towards a solution according to the update strategy previously described. A schematic of this process is provided in figure 4 above.

It has been shown that the AHN described above minimises a function containing both integer and real valued variables. This function can be set equal to the cost function of the scheduling problem. Inequality constraints may be solved by other techniques. Hence the AHN is suitable for application to the scheduling problem. A wide variety of constraints may be considered by this algorithm.

6.6 Other Neural Network Approaches To The Scheduling Problem

[21] proposed a heuristic approach to the generation scheduling problem. A multistage algorithm with two neural network stages is used. The first stage uses a

multilayer perceptron [19] to match the load forecast with a previously used template pattern. Then a schedule, previously computed with dynamic programming for this template, is extracted from a database and used as an initial schedule for the present case. The final stage also uses a multilayer perceptron to adjust the initial schedule to a more optimal one. It was found, for some examples, that this final stage failed to improve on the initial schedule. Results showed that this method compared very favourably with dynamic programming, however, no cases where the proposed algorithm outperformed dynamic programming were reported. It is possible that the good solutions found were due to a close match between load templates and forecasts, hence effectively the same algorithm was applied in each case. However, the fast computation time of this method is a significant advantage over standard dynamic programming techniques.

Kasangaki *et al.* (1995) considered the unit commitment and economic dispatch problems to be stochastic due to random variations in load and unpredictable unit outages. A stochastic Hopfield network is designed to incorporate these effects into the scheduling algorithm. However, as in the coupled gradient network [32] continuous neurons are used to represent discrete variables. The results from this method were significantly better than those obtained using a hybrid Lagrangian Relaxation - Monte Carlo algorithm.

[6] used the Hopfield network to perform security constrained corrective rescheduling of real power outputs of generators. It was found that the proposed method found solutions that were ôgenerallycomparableö to those found with the conventional linear programming based approach, but solution times were significantly longer. It was suggested that the advantage of the neural network method lay in possible parallel implementation on transputer networks leading to faster solution times.

[2] used a multi-layer perceptron to estimate parameters such as penalty factors and system losses based on generator outputs and system load. Conventional methods for estimating these parameters are quite time consuming but may be used to build a training set for the neural network. These parameters were then utilised by a conventional economic dispatch program. Results illustrated that this method preserved accuracy and gave a significant saving in computation time.

[34] applied the Hopfield network algorithm developed by [5] to the economic dispatch problem. The formulation was modified to incorporate transmission constraints. The results obtained were very close to those obtained using quadratic programming although the proposed method required less CPU time.

6.7 Conclusions

This chapter has outlined various neural network approaches to the scheduling problem. It can be seen that these techniques have matured significantly in recent years and may serve as an alternative to more conventional methods. While the methods described here have been applied to specific power system problems the general formulation allows application to many diverse applications.

6.8 References

1. Dillon J. D., 1998, "Scheduling of a hydrothermal power system" Masters thesis, Department of Electronic and Electrical Engineering, University College Dublin.
2. Djukanovi M., alovi M., Miloevi B., and Sobajic D. J., 1996, "Neural-net based real-time economic dispatch for thermal power plants", *IEEE Transactions on Energy Conversion*, Vol. 11, No. 4, pp. 755-761.
3. Fukuyama Y. and Ueki Y., 1994, "An application of neural network to dynamic dispatch using multi processors" *IEEE Transactions on Power Systems*, Vol. 9, No. 4, pp. 1759-1765.
4. Funabiki N., and Nishikawa S., 1997, "A binary Hopfield neural-network approach for satellite broadcast scheduling problems", *IEEE Transactions on Neural Networks*, Vol. 8, No. 2, pp. 441-445.
5. Gee A. and Prager R., 1995, "Limitations of neural networks for solving travelling salesman problems", *IEEE Transactions on Neural Networks*, Vol. 6, No. 1, pp. 280-282.
6. Ghosh S. and Chowdhury B. H., 1996, "Security-constrained optimal rescheduling of real power using Hopfield neural network", *IEEE transactions on Power Systems*, Vol. 11, No. 4, pp. 1743-1748.
7. Gill P. E., Murray W. and Wright M. H., 1993, "Practical Optimization", London, Academic Press.
8. Guan X., Luh P. B., Yan H. and Amalfi J. A., 1992, "An optimisation-based method for unit commitment", *Electric Power and Energy Systems*, Vol. 14, pp. 9-17.
9. Hayashi Y., Iwamoto S., Furuya S. and Liu C. C., 1996, "Efficient determination of optimal radial power system structure using Hopfield neural network with constrained noise", IEEE Transactions on Power Delivery, Vol. 11, No. 3, pp. 1529-1535.
10. Hopfield, J. J., 1982, "Neural networks and physical systems with emergent collective computational abilities", *Proceedings of the National Academy of Science*, USA, Vol. 79, pp. 2554-2558.
11. Hopfield, J. J., 1984, "Neurons with graded response have collective computational properties like those of two state neurons", *Proceedings of the National Academy of Science*, USA, Vol. 81, pp. 3088-3092.

12. Hopfield J. J. and Tank D.W., 1985, "Neural computation of decisions in optimisation problems," *Biological Cybernetics*, Vol. 52, pp. 141-152.
13. Kasangaki V. B. A., Sendaula H. M. and Biswas S.K., 1995, "Stochastic Hopfield artificial neural network for electric power production costing", *IEEE Transactions on Power systems*, Vol. 10, No. 3, pp. 1525-1533.
14. Kennedy M. P. and Chua L., 1987, "Unifying the Hopfield linear programming circuit and the canonical non-linear programming circuit of Chua and Lin", *IEEE Transactions on Circuits and Systems,* Vol. 34, No. 2, pp. 210-214.
15. King T., El-Hawary M., and El-Hawary F., 1996, "Optimal environmental dispatching of electric power systems via an improved Hopfield neural network model", *IEEE Transactions on Power Systems*, Vol. 11, No. 3, pp. 1146-1158.
16. Kumar J. and Sheblé G., 1995, "Clamped state solution of artificial neural network for real time economic dispatch," *IEEE Transactions on Power Systems*, Vol. 10, No. 2, pp. 925-931.
17. Lee F. N., Lemonidis L. and Liu K. C., 1994, "Price-based ramp-rate model for dynamic dispatch and unit commitment," *IEEE Transactions on Power Systems*, Vol. 9, No. 3, pp. 1233-1241.
18. Lee K. Y., Sode-Yome A. and Park J. H., 1997, "Adaptive Hopfield networks for economic load dispatch" to appear in *IEEE Transactions on Power Systems*, IEEE paper No. PE-217-PWRS-0-01-1997.
19. Lippmann, R. P., 1987, "An introduction to computing with neural nets", *IEEE ASSP Magazine*, April, pp. 4-22.
20. Liu Z. J., Villaseca F. E., Renovich F., 1992, "Neural networks for generation scheduling in power systems", *Proceedings of the International Joint Conference on Neural Networks,* Baltimore, MD, pp. 233-238.
21. Ouyang Z., Shahidehpour S. M., 1992b, "A multi stage intelligent system for unit commitment", *IEEE Transactions on Power Systems*, Vol. 7, No. 2, pp. 639 - 645.
22. Park J. H., Kim Y. S., Eom I. K. and Lee K. Y., 1993, "Economic load dispatch for piecewise quadratic cost function using Hopfield neural network", *IEEE Transactions on Power Systems*, Vol. 8, No. 3, pp. 1030-1038.
23. Pipes L. A. and Harvill L. R., 1985, "Applied Mathematics for engineers and physicists", McGraw-Hill.
24. Sasaki H., Watanabe M., Kubokawa J., Yorino N., and Yokoyama R, 1992, "A solution method of unit commitment by artificial neural networks", *IEEE Transactions on Power Systems*, Vol. 7, No. 3, pp. 974-981.
25. Su C. and Chiou G., 1997, "A fast computation Hopfield Method to economic dispatch of power systems", *IEEE Transactions on Power Systems*, Vol. 12, No. 4, pp. 1759-1764.

26. Tank D. and Hopfield J. J., 1986, "Simple neural optimization networks: An A/D converter, signal decision circuit, and a linear programming circuit", *IEEE Transactions on circuits and systems*, Vol. 33, No. 5, pp. 533-541.
27. Walsh M. P. and O'Malley M. J., 1997, "Augmented Hopfield network for unit commitment and economic dispatch", *IEEE Transactions on Power Systems*, Vol. 12, No. 4, pp. 1765-1774.
28. Walsh M. P., 1998, "A novel neural network for power system scheduling", Ph.D. Thesis, Department of Electrical and Electronic Engineering, University College Dublin, Ireland.
29. Walsh M. P., Flynn M. E. and O'Malley M. J., 1998, "Augmented Hopfield network for mixed integer programming", to appear in *IEEE Transactions on Neural Networks*.
30. Wang C. and Shahidehpour S., 1993, "Effects of ramp-rate limits on unit commitment and economic dispatch", *IEEE Transactions on Power Systems*, Vol. 8, No. 3, pp. 1341-1350.
31. Wang Y. and Wahl F. M., 1997, "Vector-entropy optimization-based neural-network approach to image reconstruction from projections", *IEEE Transactions on Neural Networks*, Vol. 8, No. 5, pp. 1008-1014.
32. Watta P. B. and Hassoun M. H., 1996, "A coupled gradient approach for static and temporal Mixed Integer Optimization," *IEEE Transactions on Neural Networks*, Vol. 7, No. 3, pp. 578-593.
33. Wood A. J. and Wollenberg B. F., 1996, "Power generation operation and control", Wiley.
34. Yalcinoz T. and Short M. J., 1997, "Neural networks for solving economic dispatch problem with transmission capacity constraints", to appear in *IEEE Transactions on Power Systems*, IEEE paper No. PE-807-PWRS-0-07-1997.
35. Yan H., Luh P. B., Guan X., and Rogan P. M., 1993, "Scheduling of hydrothermal power systems", *IEEE Transactions on Power Systems*, Vol. 8, pp. 1358-1365.

Chapter 7

DECISION MAKING IN A DEREGULATED POWER ENVIRONMENT BASED ON FUZZY SETS

S. M. Shahidehpour And M. I. Alomoush
Department of Electrical and Computer Engineering
Illinois Institute of Technology
Chicago, IL 60616

7.1 Fuzzy Sets: Introduction and Background

7.1.1 INTRODUCTION

In a paper written in 1965 [1], Zadeh introduced the properties of fuzzy sets, and defined it as a class of objects with a continuum of grades of membership in the interval [0,1]. This definition differs from the conventional deterministic set theory in which objects have only membership (characteristic function) values taken from the discrete set {0,1}. In fuzzy set theory, each object x in a fuzzy set X is given a membership value using a membership function denoted by $\mu(x)$ which is corresponding to the characteristic function of the crisp set whose values range between zero (complete non-membership) and one (complete membership). In fuzzy sets, the closer the value $\mu(x)$ to 1.0 the more x belongs to X.

Many objects faced in real life problems don't have precisely defined criteria of membership and can not be defined on "completely true" or "completely false" judgements. Fuzzy set comes to fill this gap as superset of conventional (Boolean) set, or it can be said that Boolean logic has been extended to fuzzy logic to handle the concept of "partial truth", where "partial truth" encompasses values between "completely true" and "completely false". Fuzzy sets are defined as functions that map a member of the set to a number between 0 and 1 indicating its actual degree of membership as means to model the uncertainty of natural language.

7.1.2 FUZZY SET THEORY VERSUS CLASSICAL SET THEORY

In the classical set theory, a subset A of a set X can be defined as a mapping from the elements of X to the elements of the set {0,1}, that is:

$$\mu_A(x) = \begin{cases} 1 & ; \quad x \in A \\ 0 & ; \quad x \notin A \end{cases} \qquad (1)$$

This mapping may be represented as a set of ordered pairs with exactly one ordered pair present for each element of X. The first element of the ordered pair is an element of the set X, and the second element is an element of the set {0,1}. The value 0 is used to represent non-membership, and the value 1 is used to represent membership. The truth or falsity of the statement "x is in A" is determined by finding the ordered pair whose first element is x. The statement is true if the second element of the ordered pair is 1, and the statement is false if it is 0. The 1 indicates membership and 0 represent non-membership. In this theory, nothing is called partial or complete membership.

Similarly, a fuzzy subset A of a set X can be defined as a set of ordered pairs, each with the first element from X, and the second element from the interval [0,1], with exactly one ordered pair present for each element of X. In other words, a subset A of a set X can be defined as a mapping from the elements of X to the elements of the interval [0,1], that is:

$$\mu_A(x) \to 0 \le \mu_A(x) \le 1; \quad x \in A \qquad (2)$$

The value 0 is used to represent complete non-membership, the value 1 is used to represent complete membership, and values in between are used to represent intermediate degrees of membership. The set X is referred to as the Universe of Discourse for the fuzzy subset A.

7.2 Fuzzy Multiple Objective Decision

Since fuzzy set theory was introduced to model uncertain environments, it may play a role in managing uncertainties that arise in a power market decision making. Zadeh and Bellman [2] proposed decision-making using the fuzzy set theory, and Saaty [3,4] introduced a new methodology for weighing the importance of decision variables. Yager [5,6] extended the approach and developed fuzzy decision making where objectives and constraints were of unequal importance.

Multiple objective decisions can be simulated mathematically and analyzed using fuzzy rules. We are concerned in this chapter with the optimization of an objective while observing constraints to fulfill our function. We consider the following cases:

7.2.1 GOAL AND CONSTRAINTS WITH EQUAL IMPORTANCE

We have a fuzzy decision problem with n_o objectives and n_c constraints, and would like the best decision to choose an alternative among X possible alternatives. Let x be an alternative such that for any criterion $\mu_i(x) \in [0,1]$ indicates the degree to which this criterion is satisfied by x. To find the degree to which x satisfies all criteria denoted by $D(x)$, we make the following statements:

- A fuzzy objective O is a fuzzy set on X characterized by its membership function:

$$\mu_O : X \to [0,1] \quad (3)$$

- A fuzzy constraint C is a fuzzy set on X characterized by its membership function:

$$\mu_C : X \to [0,1] \quad (4)$$

- Fuzzy objectives and fuzzy constraints sets must satisfy the fuzzy decision D. In this case, if goals and constraints are of equal importance, we look for x to satisfy:

$$\begin{array}{c} O_1(x) \text{ and } O_2(x) \text{ and} ... O_{n_O}(x) \text{ and} \\ C_1(x) \text{ and } C_2(x) \text{ and} ... C_{n_C}(x) \end{array} \quad (5)$$

or

$$D(x) = O_1(x) \cap O_2(x) \cap ... \cap O_{n_O}(x) \cap C_1(x) \cap C_2(x) \cap ... \cap C_{n_C}(x) \quad (6)$$

where n_C is number of constraints, n_O is number of objectives, $C_i(x)$ is fuzzy value associated with satisfaction of the ith constraints by the decision alternative x and $O_i(x)$ is fuzzy value of the ith objective for alternative x.

The fuzzy decision in this case is characterized by its membership function:

$$\mu_D(x) = \min\{\mu_O(x), \mu_C(x)\} \quad (7)$$

The best alternative x_{opt} is determined by:

$$D(x_{opt}) = \max_{x \in X} (D(x)) \quad (8)$$

or it is x_{opt} that satisfies:

$$\max_{x \in X} \mu_D(x) = \max_{x \in X} (\min\{\mu_O(x), \mu_C(x)\})$$
$$= \max_{x \in X} (\min\{\mu_{O_1}(x), ..., \mu_{O_{n_O}}(x), \mu_{C_1}(x), ..., \mu_{C_{n_C}}(x)\}) \quad (9)$$

where μ_{O_i} refers to membership function of $O_i(x)$ and μ_{C_j} refers to membership function of $C_j(x)$.

It is obvious that this equation is equivalent to finding the minimum of fuzzy objectives and constraints. The decision fuzzy set $D(x)$ defines a fuzzy intersection whose membership determines the optimal decision for alternatives $x \in X$. In this equation, both the objectives and constraints take values from the same fuzzy [0,1] space. The optimal decision alternative x_{opt} is the argument (alternative) that has the highest fuzzy degree of membership in $D(x)$ and fulfills the set of objectives and constraints. It is found that the alternative satisfies:

$$x_{opt} = \arg\{\max_{x \in X} D(x)\} \quad (10)$$

that is x_{opt} is the argument or index value when $D(x)$ achieves its maximum value.

7.2.2 OBJECTIVES AND CONSTRAINTS WITH VARYING IMPORTANCE

When objectives and constraints are of unequal importance, the previous decision-making equations should be modified such that the alternatives with high membership are more likely to be selected. One of the ideas that can perform this task is summarized as raising a fuzzy set to a positive scalar power value w_i such that the relative importance of objectives and constraints are distinguished in the net decision making fuzzy set $D(x)$ [5,6]. This can be obtained by associating higher values of w_i with more important objectives or constraints, i.e., the more important the alternative the higher the associated w_i.

If objectives and constraints are of equal importance, then the exponential weight w_i is equal to one. The modified decision making function after taking weighting into consideration will be:

$$D(x) = O^w(x) \cap C^w(x) \quad (11)$$

or

$$D(x) = \min\{O^w(x), C^w(x)\} \quad (12)$$

In this case, x_{opt} should satisfy:

$$\max_{x \in X} \mu_D^W(x) = \max_{x \in X} (\min\{\mu_o^W(x), \mu_c^W(x)\}) \qquad (13)$$

which it can be expressed as:

$$x_{opt} = \arg\{ \max_{x \in X} \mu_D^W(x)\} \qquad (14)$$

or

$$x_{opt} = \arg\{ \max_{x \in X} (\min\{ \mu_{o_1}^{w_1}, \mu_{o_2}^{w_2}, \ldots, \mu_{o_{n_o}}^{w_{n_o}}, \mu_{c_1}^{w_{n_o+1}}, \mu_{c_2}^{w_{n_o+2}}, \ldots, \mu_{c_{n_c}}^{w_{n_o+n_c}} \})\} \qquad (15)$$

7.3 Analytical Hierarchy Process

Analytic Hierarchy Process (AHP) was proposed by Satty [3,4] as a method to handle human judgment numerically in decision making processes. The Analytic Hierarchy Process is used to support complex decision problems. AHP's main foundation is the concept of priority that can be defined as the 'level of strength' of one alternative relative to another. This method assists a decision-maker to build a positive reciprocal matrix of pairwise comparison of alternatives for each criterion. A vector of priority is computed from the eigenvector of each matrix. The sum of all vectors of priorities forms a matrix of alternative evaluation. The final vector of priorities is calculated by multiplying the criteria weighted vector by the matrix of alternative evaluation. The best alternative has the highest priority value. In this proposal, a pairwise comparison matrix is constructed to evaluate the relative importance of the decision variables.

The relative importance of each objective or constraints can be obtained using paired comparison of the elements taken two at a time [3,4]. This method can be used to obtain the exponential weighting values that properly reflect the relative importance of the objective criteria and constraints entering a decision problem.

Assume we have s objects and want to construct a scale rating of these objects with respect to each other. The objects in our case are the objective criteria and constraints forming a decision problem, in this case $s = n_c + n_o$, where n_c and n_o are number of constraints and number of objectives, respectively. The decision-makers use paired comparison, i.e., comparing the objects two at a time, then specifying which object is more important. Depending on the degree of importance, values taken from 9 down to 1 should be assigned to the objects, higher values assigned to objects dominate less important objects.

For the purpose of decision making under variable importance, the paired comparison matrix **P** is formed which has the following properties:
- It is a square matrix of order s where $s = n_c + n_o$ (the sum of the number of objectives and the number of constraints).
- It has one's on the diagonal.
- $p_{ij} = 1/p_{ji}$
- The off-diagonal elements are specified by looking at the table of importance scale as given in Table 1, for example if i is weakly more important than j then $p_{ji} = 3$, and if fuzzy set i is absolutely more important than fuzzy set j then $p_{ji} = 9$, and so on.

The scale values along with their verbal interpretation are shown in Table 1. Based on the importance relations between the elements of decision-making – objectives and constraints – and based on the comparison table, we specify the elements of the comparison matrix **P**.

To compare a set of n objects (A_1, A_2, ..., A_n) in pairs according to their relative weights, the pairwise comparison matrix **P** can be written as:

TABLE 1. Pairwise Comparison Scale

Intensity of Importance	Verbal Scale
1	Equal importance of both elements
3	Weak importance of one element over another
5	Strong importance of one element over another
7	Very strong importance of one element over another
9	Absolute importance of one element over another
2,4,6,8	Intermediate values between the two adjacent judgments

$$P = [p_{ij}] = \begin{array}{c} \\ A_1 \\ A_2 \\ \vdots \\ \\ A_n \end{array} \begin{array}{c} A_1 \quad A_2 \quad \cdots \quad A_n \\ \begin{bmatrix} w_1/w_1 & w_1/w_2 & \cdots & w_1/w_n \\ w_2/w_1 & w_2/w_2 & \cdots & w_2/w_n \\ \cdot & \cdot & & \cdot \\ \cdot & \cdot & \cdots & \cdot \\ \cdot & \cdot & & \cdot \\ w_n/w_1 & w_n/w_2 & \cdots & w_n/w_n \end{bmatrix} \end{array} \quad (16)$$

where w_i/w_j refers to the ij^{th} entry of **P** which indicates how element i is compared to element j. This matrix is a reciprocal matrix ($p_{ij} = 1/p_{ji}$). In order to find the vector of weights $\mathbf{W} = [w_1 \ w_2 \ ... \ w_n]^T$, we multiply matrix $\mathbf{P}$ by the vector $\mathbf{W}$ to get:

$$\mathbf{PW} = \begin{bmatrix} w_1/w_1 & w_1/w_2 & ... & w_1/w_n \\ w_2/w_1 & w_2/w_2 & ... & w_2/w_n \\ \cdot & \cdot & & \cdot \\ \cdot & \cdot & ... & \cdot \\ \cdot & \cdot & & \cdot \\ w_n/w_1 & w_n/w_2 & ... & w_n/w_n \end{bmatrix} \begin{bmatrix} w_1 \\ w_2 \\ \cdot \\ \cdot \\ \cdot \\ w_n \end{bmatrix} = \begin{bmatrix} w_1 + w_1 + ... + w_1 \\ w_2 + w_2 + ... + w_2 \\ \cdot \\ \cdot \\ \cdot \\ w_n + w_n + ... + w_n \end{bmatrix} = n \begin{bmatrix} w_1 \\ w_2 \\ \cdot \\ \cdot \\ \cdot \\ w_n \end{bmatrix} \quad (17)$$

or
$$\mathbf{PW} = n\mathbf{W} \quad (18)$$
or
$$(\mathbf{P} - n\mathbf{I})\mathbf{W} = \mathbf{0} \quad (19)$$

The nontrivial solution of the last equation is obtained by solving this eigenvalue problem; we solve the characteristic equation $|\mathbf{P} - \lambda\mathbf{I}| = 0$, where λ represents the eigenvalue and $\mathbf{I}$ is the identity matrix. The solution gives us the eigenvalues of the comparison matrix. This last equation has a solution if and only if n is an eigenvalue of $\mathbf{P}$, i.e., it is a solution of the characteristic equation $|\mathbf{P} - \lambda\mathbf{I}| = 0$ (n= λ). Moreover, $\mathbf{P}$ has a rank of one because columns (rows) are linearly dependent on the first column (row), which concludes that all eigenvalues are zero except a nonzero eigenvalue referred as to λ_{max}. Also, sum of the eigenvalues of a positive matrix is equal to the trace of the matrix, so, we have:

$$\text{Trace } (\mathbf{P}) = \sum_{i=1}^{n} \lambda_i = n = \lambda_{max} \quad (20)$$

From this formulation, we can refer to the eigenvector corresponding to the maximum eigenvalue as $\mathbf{E}_{\lambda_{max}}$. Notice that each column of $\mathbf{P}$ is a constant multiple of $\mathbf{W}$, thus, normalizing any column of $\mathbf{P}$ gives us $\mathbf{W}$ where we will refer to the normalized vector by $\mathbf{E}_n$. Another note is that $\mathbf{P}$ is consistent, i.e., has the cardinal ratio property ($p_{ij}p_{jk} = p_{ik}$), which means if we are given any n entries with no two entries in the same column or row, we can determine the rest of entries of the matrix. If the scale in the situation under study is not known exactly, i.e., we do not know $\mathbf{W}$ but can estimate $\mathbf{P}$, then the system will no more be a consistent. However, the eigenvector is robust for changes in judgments where small errors in estimation leads to small errors in eigenvalues. In this case, our system of equations will be written as: $\mathbf{PW} = \lambda_{max}\mathbf{W}$, where $\mathbf{P}$ has a real positive eigenvalue λ_{max} close to n with a multiplicity of one, the corresponding eigenvector has nonnegative entries and the other eigenvalues are close to zero.

The estimates of weights for the compared elements can be found by normalizing the eigenvector corresponding to the largest eigenvalue. Judgments are indicated by closeness of the maximum eigenvalue and n.

After we find the priorities values w_i based on AHP, we can find the modified alternatives by raising the degrees of membership to the order of the corresponding priority value.

7.4 Applications of Fuzzy Sets

7.4.1 APPLICATION OF FUZZY SETS TO POWER SYSTEMS

Fuzzy sets are applied to different problems faced in power systems that include uncertainties in planning, scheduling, security, control, stability, load forecasting, unit commitment, and state estimation. Since heuristics, expert knowledge, experience, and intuition are essential in power system operations, fuzzy approach can be used in these areas to represent uncertainties by fuzzification of ambiguous variables and assigning membership functions based on preferences and/or experience. Relationships of uncertain variables can be represented by fuzzy rules using reasoning statements of the form " IF A THEN B". Sometimes, the fuzzy logic approach is essential to deal with a certain problem and preferred over the probability approach, which can not manage ambiguous variables. Fuzzy sets are preferred in satisfying conflicting objectives in decision problems, especially if unequal importance factors are to be taken into consideration. Some of these applications are described in the following [7-31,38-41]:

7.4.1.1 Unit Commitment
Fuzzy sets are successfully applied to unit commitment, and one of the pioneering applications in unit commitment is integrating fuzzy sets with Artificial Neural Network (ANN) to model load forecasting errors in unit commitment [18-20]. For an area in a power system, there are typical load curves corresponding to various situations. For each typical load profile, there is an optimal unit commitment schedule obtained by rigorous mathematical methods or system operation experience. An ANN can be trained for recognizing those load curves and its corresponding unit commitment solutions. A trained ANN can provide the commitment schedule for any given load curve, and even though the off-line training process is time consuming, the on-line performance of an ANN can be finalized in seconds. Since forecasting errors exist in the predicted load curves, fuzzy sets are employed to describe these uncertainties. Since the ANN output consists of fuzzy information, fuzzy set operations are performed to find the proper unit commitment for certain forecasted load profiles. Therefore, ANN enhanced by fuzzy set presents a viable option for the optimal scheduling of generating units in a power system.

7.4.1.2 State Estimation

The basic problem of state estimation is to find the estimated values that best fit the measurements. Statistical estimation models are constructed in a framework where the difference between the measurement data and estimated value are represented by the statistical observation errors. Difficulties arise when the distribution function of errors is not predicted. Recently, fuzzy sets have been employed in power system state estimation [16,17,22] where fuzzy least absolute value (LAV) estimator and fuzzy least median squares (LMS) estimator were proposed. In this application, based on the interpretation of residual measurements, a linear membership function that represents the satisfaction with respect to each measurement is developed. Then the sum of individual memberships is maximized in a global optimization.

7.4.1.3 Generator Maintenance Scheduling

The main objectives of maintenance scheduling are reducing generation cost, increasing system reliability and increasing the lifetime of units. In maintenance scheduling, some uncertainties arise in system variables, which can be dealt with fuzzy sets [41]. These uncertain variables are reserve, production cost, manpower hours, maintenance window and geographical constraints. After assigning membership functions to these uncertain variables and using fuzzy dynamic programming, the fuzzy objective and fuzzy constraints can determine the optimal generator maintenance scheduling. In this case, the optimal decision is determined by the intersection of membership functions assigned to the objective and constraints. The optimal decision has the maximum membership value at the last stage in dynamic programming.

7.4.1.4 Power System Stabilizer

Oscillations may occur, if a disturbance is applied to power systems. In this situation, power system stabilizers are essential to damp the oscillations of generator speed. To design a fuzzy stabilizer (controller), we apply some control reasoning rules of the form " *IF* A (situation) *THEN* B (action)" that represent premise (situation) and consequent (action) [23,26,39]. The input space where the rule is valid is characterized by a fuzzy set on the input variable (discourse). To implement the controller using fuzzy sets, we assign fuzzy sets to input signals of the controller (generator speed deviation and its derivative) where we refer to these inputs as "large positive", "medium positive", "small positive", "very small", "small negative", "medium negative", and "large negative". Control rules can be constructed as a fuzzy description of control logic representing the human expert's quantitative knowledge. For example, *"IF* speed deviation is low positive and its derivative is low negative, *THEN* the control action u should be very small."

7.4.1.5 Planning

To make a decision in power system expansion planning, some factors are to be taken into consideration such as demand, new station location and environmental effects. The same conditions apply to long-term and short-term scheduling, such as yearly maintenance and seasonal fuel scheduling. If these factors have ambiguous descriptions then fuzzy approach is appropriate [23,27,38].

7.4.1.6 Security

Security of power system has some vague descriptions that need a tool to take uncertainties into account. For example, when a contingency occurs, a line is overloaded because its limit is exceeded by 5%, while a line carrying 95% of its capacity is declared unaffected or safe. In addition, some lines may have a larger impact than others, depending on their locations, loads they are supplying, operating point and flows before a contingency occurs. For that reason, severity indices need to take into consideration the degree of severity, and definitions such as "highly affected" and "lightly loaded" can be used to describe the situation. To include these uncertainties, membership functions may be assigned to severity indices or to line flows in evaluating security [11-15].

7.4.1.7 Load Forecasting

Loads in power systems depend on weather conditions, economic situation, holidays, geographical locations, daylight hours, population increase and other factors. Some of these factors are uncertain but known to be within certain ranges, and most possibly will be around specific points in these ranges. The fuzzy set approach can apply reasoning and assign membership values to give more accurate load forecasts under these uncertain conditions [23,40].

7.4.1.8 Transaction Decision-Making

With deregulation in electrical utilities, many uncertainties exist in purchase transactions, sale transactions, system demand, reserve requirements and transmission charges. The right decision in the appropriate time can save money, and obtain customer satisfaction. The decision problem may be formulated as a minimization of a fuzzy non-linear objective function subject to fuzzy constraints. One scenario to solve this problem is to convert non-linear terms to linearized terms, and then use fuzzy arithmetic to formulate the problem as fuzzy linear programming. The next step is to transform the linearized fuzzy optimization problem to a linear crisp optimization and solve the problem using a linear programming algorithm [21,23,24].

7.4.1.9 Fault Diagnosis

Locating fault sections in a power system is an essential issue for system dispatchers. This is to ensure minimizing outage time, maintaining system reliability and satisfying customers. Dispatchers use heuristic rules based on the history in fault diagnosis. Fuzzy set theory is used to represent uncertainties that occur due to failures of protective relays [26], failures of breakers and errors in local acquisition and transmission. Uncertain signals are represented by membership functions to define the most likely fault sections.

7.4.1.10 Real Time Pricing of Electric Power

Bus loads in a power network are uncertain, and can be fuzzified and modeled using fuzzy sets [25]. The load may be declared as "will not be less than 50MW and not more than 60 MW, and most probably between 54 MW and 56 MW." In this application, load is modeled as a fuzzy number that has the trapezoidal membership function. The optimal power flow can use these uncertain values to solve for uncertain (fuzzy) spot price rates, and fuzzy rates can be defuzzified, using certain rules, to identify a crisp spot price.

7.4.1.11 Contingency Constrained OPF

Fuzzy logic approach was used for the contingency constrained optimal power flow (CCOPF) problem for modeling the conflicting objectives of secure operation and economic operation [31]. The problem was formulated in a decomposed form that allows for post-contingency corrective scheduling. The formulation treats the minimization of both the pre-contingency operating cost and of the post-contingency correction times as fuzzy conflicting goals. A systematic procedure for specifying the tolerance parameters that are needed to obtain fuzzy membership functions for these fuzzy goals was developed in this study. The fuzzy problem is then transformed into a crisp formulation and then solved as a conventional OPF problem. The approach used in this application yielded Pareto curves that can guide the system operator regarding the tradeoff between cost and security against contingencies.

7.4.2 FUZZY SETS APPLICATIONS IN INDUSTRY

Fuzzy logic is useful in applications where operators are not intimately familiar with all data or judgments must be synthesized from multiple input. Fuzzy logic has been successfully applied to information retrieval systems, a navigation system for automatic cars, a predicative fuzzy logic controller for automatic operations of trains, laboratory water level controllers, robot arc-welders controllers, robot vision controllers, speech recognition, image recognition, drives and machines and many other applications. Some applications are summarized in the following [32-34]:

7.4.2.1 Cement kiln control

It was the first application of fuzzy logic in industry in Denmark in 1970 for controlling cement kiln processes [33]. The operator was required to monitor, control and manage internal states of kiln and operations of a large number of rules of thumb that relate internal states of kiln, chemical interactions and control operations. One of these rules is "If the oxygen percentage is rather high and the free-lime and kiln-drive torque rate is normal, decrease the flow of gas and slightly reduce the fuel rate."

7.4.2.2 Blast Furnace Control

Blast furnaces are essential parts in iron and steal processes, and their thermal conditions used to be controlled based on operators' experience and knowledge [32]. When the complexity and size of furnace increase, and when accurate measurement of internal thermal conditions is difficult due to unreliable sensors, then the data that describe thermal condition are uncertain. The fuzzy set approach is used to monitor and control the process by assigning membership functions as a means of expressing the fuzziness associated with the blast furnace. Inside the furnace there are high temperature, high pressure and extremely complex reactions in gas, solid and liquid phases. The blast furnace is required to respond swiftly and flexibly to changes in production plans and operational changes and to produce high quality molten iron. The thermal condition is controlled by the quantity of raw material as well as the state of blast, temperature and moisture. Based on operators' experience, thermal conditions is divided into seven levels given as "very low", "low", "somewhat low", "normal", "somewhat high", "high" and "very high", and for each of these expressions a corresponding safety factor is specified. Three-dimensional membership functions for hot metal temperature, thermal level and certainty factors are developed to control furnace processes. The effect of fuzzy control is summarized as simplification of knowledge expressions, reduction in number of rules, and improvement in real time control.

7.4.2.3 Cold Rolling Control

Cold rolling is a process for rolling strip to prescribed thickness and shapes at low temperatures [32]. Initial set-up control provides the basis for calculating roll gap, rolling speed, and the tensile force applied to the material being rolled. This control is the most important part that defines productivity, stability and quality of the product. In this process, friction between the roll and rolled materials and deterioration resistance of the materials are hypothetically figured and used in control equations. Therefore, large errors are introduced in the results that are corrected by adjusting the set-up control values. These adjustments may result in a reduced rolling speed, which in tern affect productivity. For these reasons fuzzy logic based reasoning is used to replace the operators' experience-based control, which is integrated with the set-up model. The fuzzy model is used to predict the rolling load which is unknown to operators. The inputs to the fuzzy model are rolled material information, desired width and thickness, tension information, and the output is the rolling load. A learning model is integrated with the fuzzy model to correct the rules of the rolling load prediction model. The rolling load prediction model expresses rolling loads, its output, as sets of " IF X_1 is A_1, X_2 is A_2,...,X_m is A_m, THEN $Y=a_0+a_1 X_1+....+a_m X_m$."

7.4.2.4 Automatic Speed Control

In the conventional speed cruise control methods, control devices assume fixed dynamic characteristics and the disturbance is small, while the dynamic characteristics vary with gear changes, load and road conditions [32]. Fuzzy logic controller, constructed of qualitative relations between the car's carburetor opening and speed, handle changes in the dynamic characteristics of the car. The inputs to the fuzzy logic controller are based on speed deviations. One of the fuzzy linguistic control rules used in the controller is " If the speed is below the target speed, press down the accelerator." Using fuzzy logic controller provides smooth acceleration unaffected by gear changes and stable operation unaffected by road's declines and inclines, and in general it is effective for varying dynamic characteristics of the car.

7.4.2.5 Rainwater Pump Management

Rainwater pumps are used to remove rainwater to rivers or waterways after collected by sewer lines, and their operation should take into consideration changes in the amount of water and other factors such as pump well level and number of pumps [32]. Pump facilities, which cannot be fully automated, have been managed by experienced operators who take into consideration well levels, rainfall conditions, past operational experience and other factors. Fuzzy adaptive control that makes adaptations to the changes in plant parameters is used for pump management to replace experienced operators using control rules. Control rules include circumstantial judgments of the control object and operational methods using "IF A THEN B" form, where A is antecedent and B is consequent. One of these rules is " When the present pump well level is high and the forecast is for further increase, one more pump should be activated." The variables in antecedent are water level, water level change, pump output and rainfall intensity and the variables in consequent are pump output change and rainwater inflow condition. All these variables represented by membership functions for their fuzzy expressions as "low", "somewhat low", "close to", "large reduction", "small increase" and other terms. Using adaptive control will result in easiness of control rule design, reduction in control rules, and increase in robustness of control.

7.4.2.6 Automatic Train Operation

The aims of control system in a train operation are accelerating the train at a departure signal, regulating train's speed and stopping the train at a fixed position at the next station. Experienced operators by maintaining these aims are in turn looking for the following objectives: safety, comfort, energy saving, speed, elapsed time, and stopping precision. These objectives are considered as indices and defined by fuzzy sets and expressed in linguistic variables such as "good", "bad", "very good" , "intermediate", and so on [32]. Predictive fuzzy control is used in train operation in place of PID (proportional-integral-deferential) controllers.

7.4.2.7 Fuzzy Environmental Control

A precision temperature and humidity controller with a fuzzy control unit [34] was designed to control the environment for large computing systems. The systems themselves generate heat and are specified by their manufacturers to be maintained in as little as a plus-or-minus 1 degree range. Humidity causes corrosion and jamming of associated mechanical systems at high humidity levels and the enhanced possibility of static discharge at lower levels. Environmental control system faces nonlinearities, caused by such system behavior. Uncertainties in system parameters are often present, for example, room size and shape, location of heat-producing equipment, thermal mass of equipment and walls, and amount and timing of external air introduction. A fuzzy logic control system has six fuzzy inputs, three fuzzy outputs, and 144 principles (rules). Fuzzy input variables are: the temperature relative to a setpoint, the rate of temperature change, the humidity relative to a setpoint, the rate of humidity change, and two proprietary variables associated with the action of the controllers. Fuzzy outputs control amount of cooling, amount of dehumidification, and heat. Each fuzzy variable is assigned seven membership functions as values, with the traditional "large negative", "medium negative", "small negative", "near zero", "small positive", "medium positive", and "large positive" as labels. Examples of a temperature control principle, using the "AS ...THEN" are: "AS temperature relative to set point is "small positive" and temperature rate of change is "medium positive" THEN amount of cooling is "small positive"., "AS temperature relative to setpoint is "small negative" and amount of cooling is "small positive" THEN wait delay to cooling change is "medium positive." A fuzzy OR operator (maximizer) is used as the defuzzification technique.

7.5 Decision Making in a Deregulated Power Environment Based on Fuzzy Sets

Electric power companies face many uncertainties in their daily scheduling of inter-utility transactions. With the deregulation of electric utilities, the competition in electricity market will introduce additional uncertainties in transactions, system demands, reserve requirements and prices of future purchases. These uncertainties should be managed properly in order to offer right decisions in an appropriate time. In this chapter, a decision support approach is described to facilitate the selection of best offer among many alternatives. The system variables are MW demand requirements, transmission charges, congestion risk, ancillary services and generation availability. Decision making in deregulated power environment with unequal importance factors is also introduced. The chapter introduces examples for the application of fuzzy rules in decision making.

In deregulated open access and competitive power environment, Distribution Companies (DisCos), Generation Companies (GenCos), Independent Power Supplies (IPSs), Independent System Operator (ISO) and other power market's entities face several types of uncertainties in system demand, future transacted power, reserves, fuel prices, available transmission capacity, unit availability, incremental and decremental prices, congestion risk, transmission charges, available bids and others. The effects of these uncertainties, if not managed properly, may significantly affect the economics of power market and lead to unreliable system, which costs all parties time, money and confusion. In such an uncertain environment, the decision to be taken should be reasonable, economical and not risky.

Linguistic expressions and experience can describe uncertain variables in an accurate description for these variables. In this context, Analytic Hierarchical Process (AHP) can play an important role to differentiate between the contributions of these variables to a decision. The statements "around", "essentially", "not more than", "somewhere between" and other ambiguous expressions express the situation in the power market. The ambiguous expressions are an essential part of the power market description. This description has a fuzzy nature, and the crisp (deterministic) conventional methods that deal with exact description may not be able to make a certain party takes the best decision in a short time. A DisCo, for example, may lose a better transaction opportunity with a GenCo or an IPS because of finalizing an improper contract with the other party without taking into consideration the future transactions uncertainties.

Fuzzy set was introduced in power system for the first time to support the long-range decision making problem [27], and later was applied to planning, operation and control of power systems [7-31,38-41] for representing uncertainties. For these applications, fuzzy sets are applied for system scheduling and inter-utility transaction, and fuzzy dynamic programming has been developed to obtain the commitment and dispatch of thermal and hydro units, and for multi-area scheduling with fuzzy demand and tie capacity limit and for power system scheduling with fuzzy reserve requirements, also, fuzzy sets are successfully applied to power flow algorithms and for optimal power flow as well.

In this chapter, a decision support approach is described to help a DisCo select the best offer among many alternatives to provide the required MW capacity and reserves. The system variables are MW demand requirements, transmission charges, congestion risk, ancillary services and generation availability. Decision making in deregulated power environment with unequal importance factors is also introduced, where a fuzzy decision tool is used to include the effects of system priorities on decision processes.

7.5.1 NOMENCLATURE

ANS_i Ancillary reserve corresponding to $GenCo_i$, $DisCo_i$, or IPS_i

$D(x)$ Decision fuzzy set

DisCo	Distribution company
GenCo	Generation company
IPS	Independent Power Supplier
CA_i	MW capacity corresponding to $GenCo_i$, $DisCo_i$, or IPS_i
n_d	Number of distribution companies
n_g	Number of generation companies
n_i	Number of independent power providers
P	Pairewise comparison matrix
p_{ij}	ij^{th} element in the parewise comparison matrix
s	Order of the comparison matrix

7.5.2 TRANSMISSION PRICING

There are some traditional transmission charging methods, namely, the postage stamp method, the contract path method and the MW-Mile method. Other methods are spot price or incremental price based, such as the marginal pricing method and the embedded cost method. We will briefly review the MW-Mile method that is used in this chapter, and the reader may refer to [35,36] for more details. Another issue that is directly related to power transaction between two parties is the transmission losses caused by that transaction. Transmission charges may include extra charges for the loss consideration.

- Power Flow MW-Mile charge: It takes into account the effect of each transaction on the flow distribution on transmission lines. It is based on line flows due to a transaction and on lengths of lines. If P_j^i is the loading of line j due to transaction i, D_j is the length of line j and R_j is the required revenue per unit length of line j, the MW-Mile method uses the following equation to find the MW-Mile (electrical distance) charge C_d^i corresponding to transaction i:

$$C_d^i = \sum_{j=1}^{n} \frac{P_j^i \, D_j \, R_j}{\sum_{transactions} P_j^i} \qquad (21)$$

- Loss-Included charges: For each transaction between a DisCo and a GenCo, there is a corresponding power transmission loss. The share of each transaction is specified, so the corresponding charge is known. ISO may broadcast an n-dimensional loss vector for n nodes of the transmission network

[37]. From this vector, losses can be estimated once injection point, extraction point and transaction magnitude in MW are known. If we denote this vector by L, injection node by k and extraction point by m, then for a transaction of P_t MW, the estimated loss is $P_t(L(k) - L(m))$. Depending on this loss, appropriate charge can be defined.

7.5.3 DESCRIPTION OF THE PROBLEM

A decision is to be taken by a Decision-Making Committee (DMC) in a Distribution Power Company (DisCo). The DMC is trying to solicit an appropriate offer in the open power market among alternatives offered by generating companies (GenCos), Distribution Companies (DisCos), Independent Power Suppliers (IPSs). Let's assume that $DisCo_1$ requests a purchase of MW capacity (CA) and ancillary services (ANS). Also, suppose that some of GenCos, DisCos, or IPSs are able to provide the requested power at different prices, availability, transmission charges, and congestion developing risk, all subjected to uncertainty in data. The desired offer should be of minimum cost. The offers are evaluated via a set of performance factors for transmission charges, availability, and congestion risk. Figure 1 illustrates the available GenCos, DisCos or IPSs to supply $DisCo_1$.

Figure 2 shows the procedure for selecting the proper GenCo, IPS, or DisCos where DisCos pass the information to DMCs to buy power from the open power market. For the requested MW capacity, let us express the problem as follows:

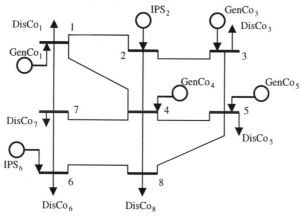

Figure 1. Eight-bus system illustrates selection of a provider.

What is the best offer among available alternatives in the open power market which provides the MW demand and reserves subject to low transmission charges, high availability and low congestion risks ?

For this situation, as shown in Figure 2, each $DisCo_i$ passes information to its DMC_i regarding its desire to buy the best offer(s) of MW capacity and reserves from the power market. Decision makers express goals and constraints required by the DisCo by fuzzy sets and curves representing DMCs satisfaction that should be matched by the desired selected alternative(s). These satisfaction measures include membership functions for each objective or constraint DisCo likes to satisfy. Based on the knowledge acquired from the market, DMCs will know market participants – DisCos, GenCos, or IPSs – who can provide the required MW capacity and/or reserves. This can be based on bids provided to power exchanges (PXs) and schedules coordinators (SCs), information obtained directly from other parties, as well as experience with monitoring transactions and market movements. All participants' offers will be evaluated under the satisfaction measures issued by DMCs which reflect DiScos' desires. Each DMC_i finalizes its decision and passes back the decision to its DisCo.

Decision-makers will base their decisions on fuzzy individual decision making approach to match the DisCo demand and ancillary services.

7.5.4 BEST OFFER DECISION

7.5.4.1 Best Capacity Offer (Capacity Decision Making)

Suppose that DMC_1 found that any of $GenCo_1$, IPS_2, $GenCo_3$, $GenCo_4$, and $GenCo_5$ was able to provide the required purchase passed by $DisCo_1$. The prices offered by these parties ($G_0(CA_i)$) and the membership of each price ($\mu_0(CA_i)$) are shown in Table 2. Figure 3 shows the membership function of the price by which all alternatives will be evaluated. For this situation:

$O = \{(CA_1, 0.78), (CA_2, 0.75), (CA_3, 0.86), (CA_4, 0.83), (CA_5, 0.80)\}$

The first constraint requires a lower risk capacity that is computed based on forced outages of generating units. Table 3 shows the generation availability data for all alternatives ($G_1(CA_i)$) and the membership of each alternative ($\mu_1(CA_i)$), and Figure 4 shows an illustration of the data under study. For this case:

$C_1 = \{(CA_1, 0.85), (CA_2, 0.91), (CA_3, 0.79), (CA_4, 0.90), (CA_5, 0.87)\}$

TABLE 2. GenCos Offered Prices and Memberships

Offered MW Capacity	Provider	$G_0(CA_i)$ $/MWh	$\mu_0(CA_i)$
CA_1	$GenCo_1$	7.0	0.78
CA_2	IPS_2	7.5	0.75
CA_3	$GenCo_3$	5.5	0.86
CA_4	$GenCo_4$	6.0	0.83
CA_5	$GenCo_5$	6.5	0.80

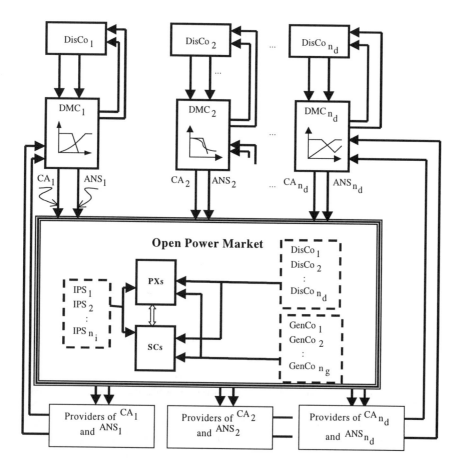

Figure 2. Power purchase procedure

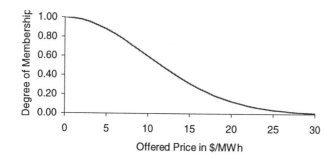

Figure 3. Offered Prices and Memberships

TABLE 3. Availability Data and Memberships

Offered MWCapacity	$G_1(CA_i)$ Outage Rate %	$\mu_1(CA_i)$
CA_1	1.40	0.85
CA_2	1.10	0.91
CA_3	1.70	0.79
CA_4	1.15	0.90
CA_5	1.30	0.87

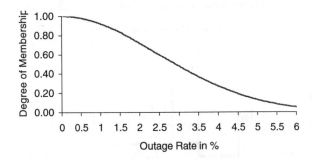

Figure 4. Membership of Generation Availability

- The second constraint considers transmission charges for different providers. We assume two separate charges: power flow MW-Mile charges and loss-included charges. MW-Mile charges ($G_{2,d}(CA_i)$) and membership values ($\mu_{2,d}(CA_i)$) are shown in Table 4. Figure 5 shows the membership function of MW-Mile. Power loss charges ($G_{2,L}(CA_i)$) and their memberships ($\mu_{2,L}(CA_i)$) are shown in Table 5 and membership function of $C_{2,L}$ is shown in Figure 6. Hence, transmission charges can be represented by the following:

$$C_{2,d} = \{(CA_1, 0.90), (CA_2, 0.81), (CA_3, 0.90), (CA_4, 0.77), (CA_5, 0.87)\}$$
$$C_{2,L} = \{(CA_1, 1.00), (CA_2, 0.87), (CA_3, 1.00), (CA_4, 0.75), (CA_5, 0.81)\}$$

TABLE 4. MW-Mile Charges and Memberships

Offered MW Capacity	$G_{2,d}(CA_i)$ $/MWh	$\mu_{2,d}(CA_i)$
CA_1	1.2	0.90
CA_2	1.7	0.81
CA_3	1.2	0.90
CA_4	1.9	0.77
CA_5	1.4	0.87

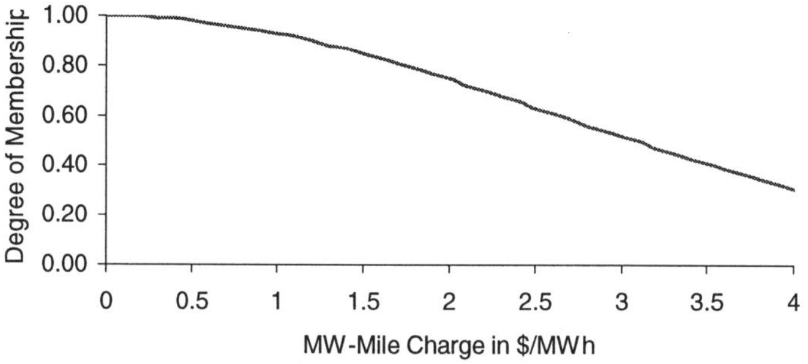

Figure 5. MW-Mile Charges

TABLE 5. Power Loss Charges

Offered MW Capacity	$G_{2,L}(CA_i)$ $/MWh	$\mu_{2,L}(CA_i)$
CA_1	3.50	1.00
CA_2	5.43	0.87
CA_3	2.50	1.00
CA_4	6.75	0.75
CA_5	6.10	0.81

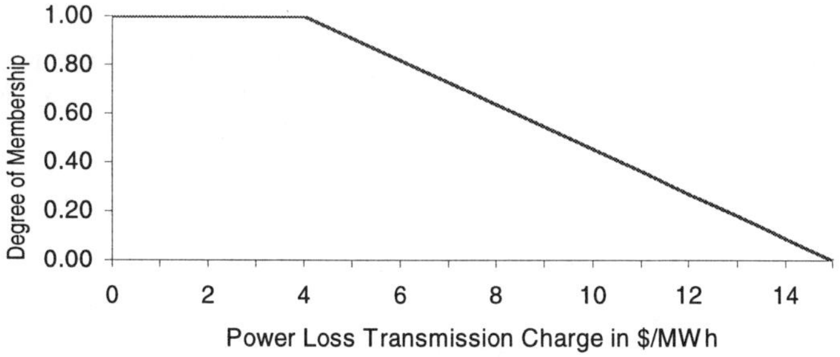

Figure 6. Power Loss Transmission Charges

The combined transmission charges are the intersection of two fuzzy sets, i.e.:

$$C_2 = C_{2,d} \cap C_{2,L}$$

$$C_2 = \min \begin{Bmatrix} (CA_1, 0.90), (CA_2, 0.81), (CA_3, 0.90), (CA_4, 0.77), (CA_5, 0.87) \\ (CA_1, 1.00), (CA_2, 0.87), (CA_3, 1.00), (CA_4, 0.75), (CA_5, 0.81) \end{Bmatrix}$$

or

$$C_2 = \{(CA_1, 0.90), (CA_2, 0.81), (CA_3, 0.90), (CA_4, 0.75), (CA_5, 0.81)\}$$

- The third constraint is based on the developed congestion. Some contracts with different providers may lead to line congestion or some contracts may cause more congestion than other contracts. DisCos should look for less risky contracts. Recently, some approaches have proposed zones for congestion management purposes, where zones' boundaries correspond to frequently congested lines, and when new lines become congested in real time, new boundaries are defined. DisCos should try to choose the offers such that the possibility of causing congestion is minimized. We don't know priory what will be the situation exactly in the real time, so we base our decision on lines with congestion possibilities. These lines are specified and declared by ISO as congestion lines. To include congestion in our decision-making function, let $p_{k,i}$ refers to flow in the predefined congested line k due to a transaction of P_i MW with provider i, and S_c is the set of congestion lines, then comparisons between available offers – from congestion point of view – can be done using the Congestion Index CI_i with the following form:

$$CI_i = \frac{\sum_{k \in S_c} p_{k,i}}{P_i} \qquad (22)$$

For each offer i, and by knowing magnitude of power to be transacted (P_i MW), lines parameters (resistance and reactance) and point of injection (provider's point) and point of extraction (DisCo's point), flows in all lines can be calculated using an approximate load flow method. After that, DisCos can find flows in the declared congestion lines corresponding to each transaction, and find congestion indices using Equation 21. For example, for a transaction of 50 MW with GenCo$_3$ ($P_3 = 50MW$), a 20% value for CI_3 means that 20% of 50 MW will flow in the predefined frequently congested lines. Each DisCo should try to minimize its total flow on declared congestion lines due to its transaction with a provider. If total power flow in these congestion lines is large, DisCo will be more involved in congestion relief issues such as congestion rents, re-dispatching contracts by ISO and using FACTS devices and phase shifters. The congestion data for offers ($G_3(CA_i)$) and membership values ($\mu_3(CA_i)$) are shown in Table 6 with membership function in Figure 7. From this information, we formulate the third constraint C_3 as:

$$C_3 = \{(CA_1, 0.76), (CA_2, 0.82), (CA_3, 0.85), (CA_4, 0.84), (CA_5, 0.89)\}$$

TABLE 6. Flow in Congested Lines

Offered MW Capacity	$G_3(CA_i)$ CI_i in %	$\mu_3(CA_i)$
CA_1	16.0	0.76
CA_2	13.0	0.82
CA_3	12.2	0.85
CA_4	12.0	0.84
CA_5	10.0	0.89

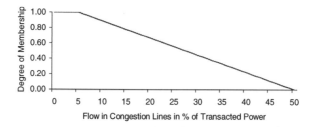

Figure 7. Possible Flow Congestion

Case I: Objective and constraints have equal importance
In this case, using fuzzy objective and constraints formulated before, the decision function for the best offer is:

$$D(x) = \min \begin{cases} (CA_1, 0.78), (CA_2, 0.75), (CA_3, 0.86), (CA_4, 0.83), (CA_5, 0.80) \\ (CA_1, 0.85), (CA_2, 0.91), (CA_3, 0.79), (CA_4, 0.90), (CA_5, 0.87) \\ (CA_1, 0.90), (CA_2, 0.81), (CA_3, 0.90), (CA_4, 0.75), (CA_5, 0.81) \\ (CA_1, 0.76), (CA_2, 0.82), (CA_3, 0.85), (CA_4, 0.84), (CA_5, 0.89) \end{cases}$$

or

$$D(x) = \{(CA_1, 0.76), (CA_2, 0.75), (CA_3, 0.79), (CA_4, 0.75), (CA_5, 0.80)\}$$

The best capacity offer is that of $GenCo_5$. Now, let us apply the comparison matrix scheme to find the best offer where all elements of **P** are ones. In this case:

$$\lambda_{max} = 4.0$$
$$E_{\lambda_{max}} = [1 \ 1 \ 1 \ 1]^T$$
$$W = E_n = [0.25 \ 0.25 \ 0.25 \ 0.25]^T$$

or all fuzzy values have the same importance weights which leads to the results of the case in the last paragraph. Using the equal weights obtained, the decision function is:

$$D(x) = \min \begin{cases} (CA_1, 0.9398), (CA_2, 0.9306), (CA_3, 0.9630), (CA_4, 0.9545), (CA_5, 0.9457) \\ (CA_1, 0.9602), (CA_2, 0.9767), (CA_3, 0.9428), (CA_4, 0.9740), (CA_5, 0.9658) \\ (CA_1, 0.9740), (CA_2, 0.9487), (CA_3, 0.9740), (CA_4, 0.9306), (CA_5, 0.9487) \\ (CA_1, 0.9337), (CA_2, 0.9516), (CA_3, 0.9602), (CA_4, 0.9573), (CA_5, 0.9713) \end{cases}$$

or

$$D(x) = \{(CA_1, 0.9337), (CA_2, 0.9306), (CA_3, 0.9428), (CA_4, 0.9306), (CA_5, 0.9487)\}$$

which again concludes that $GenCo_5$ is the best offer.

Case II: Objective and the constraints are not of equal importance

For the case of fuzzy sets with variable importance, let's study the following power market problem:

The DMC is trying to choose the appropriate offer among alternatives in the open power market. The Power Company requests a purchase of MW capacity. The desired offer should have the minimum cost with low transmission charges and congestion risk, and high availability. Moreover, availability is slightly more important than transmission charges, and transmission charges are highly more important than congestion risk. The MW price is of absolute importance as compared to transmission charges, availability and congestion considerations.

Using the importance scale in Table 1, 'The MW price is of absolute importance as compared to transmission charges, availability and congestion" means that:

$$p_{12} = 9, \ p_{21} = 1/9$$
$$p_{13} = 9, \ p_{31} = 1/9$$
$$p_{14} = 9, \ p_{41} = 1/9$$

The statement *"availability is slightly more important than transmission charge"* means: $p_{23} = 3$, $p_{32} = 1/3$. The statement *"transmission charges are highly more important than congestion risk"* means that: $p_{34} = 5$, $p_{43} = 1/5$. In this case:

$$P = \begin{bmatrix} 1 & 9 & 9 & 9 \\ 1/9 & 1 & 3 & 3 \\ 1/9 & 1/3 & 1 & 5 \\ 1/9 & 1/3 & 1/5 & 1 \end{bmatrix}$$

$$\lambda_{max} = 4.4956$$

$$W = E_n = [0.9713 \quad 0.1909 \quad 0.1299 \quad 0.0565]^T$$

by using (16), the decision will be based on the following function:

$$D(x) = \min\{O^W(x), C^W(x)\} =$$

$$\min \begin{cases} (CA_1, 0.786), (CA_2, 0.756), (CA_3, 0.864), (CA_4, 0.834), (CA_5, 0.805) \\ (CA_1, 0.969), (CA_2, 0.982), (CA_3, 0.956), (CA_4, 0.980), (CA_5, 0.974) \\ (CA_1, 0.986), (CA_2, 0.973), (CA_3, 0.986), (CA_4, 0.963), (CA_5, 0.973) \\ (CA_1, 0.985), (CA_2, 0.989), (CA_3, 0.991), (CA_4, 0.990), (CA_5, 0.993) \end{cases}$$

or

$$D(x) = \{(CA_1, 0.786), (CA_2, 0.756), (CA_3, 0.864), (CA_4, 0.834), (CA_5, 0.805)\}$$

The best offer is that of $GenCo_3$. For other possible importance relations, the offer may change.

7.5.4.2 Best Ancillary Services Offer (Ancillary Services Decision Making)

Ancillary services can be purchased separately or can be purchased along with the MW capacity. Ancillary services may be purchased from a party different from that of the MW capacity. In any case, decision-making process for ancillary service is the same as that of the MW capacity. If the objective is the cost, then the decision-makers use curves for cost similar to that of Figure 3. If the objective is on reserve requirements, then a membership function is needed for the reserve. Let's consider the following situation:

"A distribution company is looking for a provider in the power market that can provide the required ancillary services (reserve). The reserve demand is to be around 30 MW, not more than 40 MW and not less than 20 MW. The reserve must have high availability, low transmission charges and low congestion risk.

Moreover, from the point of view of DisCo, the amount of reserve is strongly more important than its availability and slightly more important than both transmission charges and congestion risk. On the other hand availability, transmission charges and congestion risk are of equal importance."

The objective is to choose the best offer among ancillary service providers (GenCos, or DisCos or IPSs) that can cover ancillary services with high availability, minimum transmission charges constraints, and minimum congestion risk.

The parties in the power market that can provide this service are: $GenCo_1$, IPS_2, $DisCo_3$, IPS_6 and $DisCo_8$. Table 7 shows reserves with a membership function in Figure 8.

TABLE 7. Reserves and its Membership

Reserve	Provider	$G_4(ANS_i)$ Reserve in MW	$\mu_4(ANS_i)$
ANS_1	$GenCo_1$	29.0	0.90
ANS_2	IPS_2	28.5	0.85
ANS_3	$DisCo_3$	27.5	0.75
ANS_6	IPS_6	30.0	1.00
ANS_8	$DisCo_8$	28.0	0.80

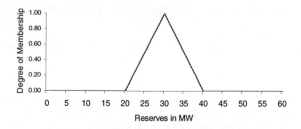

Figure 8. Reserve and its Membership

- The objective:

$O = \{(ANS_1, 0.90), (ANS_2, 0.85), (ANS_3, 0.75), (ANS_6, 1.0), (ANS_8, 0.80)\}$

- The constraints can be formulated as before. Let's assume that:

$C_1 = \{(ANS_1, 0.85), (ANS_2, 0.91), (ANS_3, 0.60), (ANS_6, 0.90), (ANS_8, 0.85)\}$

$C_2 = \{(ANS_1, 0.90), (ANS_2, 0.78), (ANS_3, 0.90), (ANS_6, 0.90), (ANS_8, 0.75)\}$

$C_3 = \{(ANS_1, 1.00), (ANS_2, 1.00), (ANS_3, 1.00), (ANS_6, 1.00), (ANS_8, 1.00)\}$

The comparison matrix is formed as:

$$\mathbf{P} = \begin{bmatrix} 1 & 7 & 3 & 3 \\ 1/7 & 1 & 1 & 1 \\ 1/3 & 1 & 1 & 1 \\ 1/3 & 1 & 1 & 1 \end{bmatrix}$$

Using the same procedure as before, we find that:

$$\mathbf{W} = \mathbf{E}_n = [0.9186 \quad 0.1987 \quad 0.2415 \quad 0.2145]^T$$

In this case the decision will use the following function:

$$D(x) = \min\{O^W(x), C^W(x)\} =$$

$$\min \begin{Bmatrix} (ANS_1, 0.908), (ANS_2, 0.861), (ANS_3, 0.768), (ANS_6, 1.000), (ANS_8, 0.815) \\ (ANS_1, 0.968), (ANS_2, 0.981), (ANS_3, 0.947), (ANS_6, 0.979), (ANS_8, 0.968) \\ (ANS_1, 0.975), (ANS_2, 0.942), (ANS_3, 0.975), (ANS_6, 0.975), (ANS_8, 0.933) \\ (ANS_1, 1.000), (ANS_2, 1.000), (ANS_3, 1.000), (ANS_6, 1.000), (ANS_8, 1.000) \end{Bmatrix}$$

or

$D(x) = \{(ANS_1, 0.908), (ANS_2, 0.861), (ANS_3, 0.768), (ANS_6, 0.975), (ANS_8, 0.815)\}$

The best offer is that of IPS_6.

7.5.5 CONCLUSION

Power Companies face many uncertainties in their daily transaction scheduling. With the deregulation of electric utilities, the competition in the electricity market will increase and cause additional uncertainties in transactions, system demands, reserve requirements and prices. Decision-makers in power markets will try to answer the questions: which is the best offer of supplying the MW needed for the demands among the available many offers? What is the best decision of fuzzy offers should be selected keeping the fuzzy constraints not violated? What will be best ancillary service subject to variable importance uncertain constraints?, etc. For the purpose of dealing with such a vague environment, a decision-making methodology has been introduced in this chapter using fuzzy sets. The chapter shows how to use fuzzy tools to deal with equal importance and variable importance fuzzy sets. This study may be extended to encompass other uncertain variables for the purposes of decision making in a deregulated environment.

7.6 References

1. Zadeh, L.A. (1965) Fuzzy Sets, *Information and Control*, 8, 338-353.
2. Bellman, R.E. and Zadeh, L.A. (1970) Decision Making in a Fuzzy Environment, *Management Science*, 17, 141-164.
3. Saaty, T.L. (1977) A Scaling Method for Priorities in Hierarchical Structures, *Jour. Math. Psychology*, 15, 234-281.
4. Saaty, T.L. (1980) The Analytic Hierarchy Process: Planning, Priority Setting, Resource Allocation, Graw-Hill, NY.
5. Yager, R.R. (1979) Multiple Objective Decision-Making Using Fuzzy Sets, *Intl. Jour. Man Machine Studies*, 9, 1375-382.
6. Yager, R.R. (1981) A New Methodology for Ordinal Multi-Objective Decisions Based on Fuzzy Sets, *Decision Sciences*, 12:4, 589-600.
7. Abdul-Rahman, K. and Shahidehpour, M. (1994) Application of Fuzzy Sets to Optimal Reactive Power Planning with Security Constraints, *IEEE Transactions on Power Systems*, 9, No. 2, 589-597.
8. Abdul-Rahman, K. and Shahidehpour, M. (1994) Reactive Power Optimization using Fuzzy Load Representation, *IEEE Transactions on Power Systems*, 9, No. 2, 898-905.
9. Wang, C., Shahidehpour, M. and Adapa, R. (1992) A Fuzzy Set Approach to Tie Flow Computation in Optimal Power Scheduling, *Proc. Of the 54th Annual Meeting of the American Power Conference*, Chicago, IL, 1443-1450.
10. Shahidehpour, M. (1991) A Fuzzy Set Approach to Heuristic Power Generation Scheduling with Uncertain Data, *Proceedings of 1991 NSF/EEI Workshop*, Norman, OK.
11. Abdul-Rahman, K. and Shahidehpour, M. (1993) Optimal Reactive Power Dispatch with Fuzzy Variables, *Proceedings of 1993 IEEE International Symposium on Circuits and Systems*, Chicago, IL.
12. Abdul-Rahman, K. and Shahidehpour, M., (1993) Application of Fuzzy Sets to Optimal Reactive Power Planning with Security Constraints, *Proceedings of the 1993 Power Industry Computer Applications (PICA) Conference*, Phoenix, AZ.

13. Abdul-Rahman, K. and Shahidehpour, M. (1993) A Fuzzy-Based Optimal Reactive Power Control, *IEEE Transactions on Power Systems*, 8, No. 2, 662-670.
14. Abdul-Rahman, K. and Shahidehpour, M. and Daneshdoost, M. (1995) AI Approach to Optimal Var Control with Fuzzy Reactive Loads, *IEEE Transactions on Power Systems*, 10, No. 1, 88-97.
15. Abdul-Rahman, K. and Shahidehpour, M. (1995) Static Security in Power System Operation with Fuzzy Real Load Conditions, *IEEE Transactions on Power Systems*, 10, No. 1, 77-87
16. Labudda, K. and Shahidehpour, M. (1993) A Fuzzy Multi-Objective Approach to Power System State Estimation, *Proceedings of International Conference on Expert System Applications to Power Systems*, Melbourne, Australia.
17. Labudda, K. and Shahidehpour, M. (1994) A Fuzzy Multi-Objective Approach to Power System State Estimation, *International Journal of Engineering Intelligent Systems*, 2, No. 2, 83-92.
18. Wang, C. and Shahidehpour, M. (1991) Application of Neural Networks in Generation Scheduling with Fuzzy Data, *Proceedings of 1991 American Power Conference*, Chicago, IL.
19. Wang, C. and Shahidehpour, M. (1992) A Fuzzy Artificial Neural Network for Multi-Area Optimal Power Generation Scheduling with Transmission Losses, *Proceedings of 1992 American Power Conference*, Chicago, IL.
20. Wang, C. (1992) Fuzzy Artificial Intelligence Approaches to Optimal Generation Scheduling of large-scale Hydro-Thermal Power Systems with Uncertain Data, Ph.D. Thesis, Department of Electrical and Computer Engineering, Illinois Institute of Technology, Chicago, IL.
21. Ferrero, R. and Shahidehpour, M. (1997) Short-term Power Purchases Considering Uncertain Prices, *IEE Proc.*, 144, No. 5, 423-428.
22. Marwali, M. (1994) Fuzzy Least Median Square Estimator in Power Systems, M.S. Thesis, Department of Electrical and Computer Engineering, Illinois Institute of Technology, Chicago, IL.
23. Momoh, J.A., Ma, X.W. and Tomsovic, K. (1995) Overview and Litrature Survey of Fuzzy Set Theory in Power Systems, *IEEE Trans. on Power Systems*, 10, No. 3, 1676-1690.
24. Yan, H. and Luh, P. (1997) A Fuzzy Optimization-Based Method for Integrated Power System Scheduling and Inter-Utility Power Transaction with Uncertainties, *IEEE Transactions on Power Systems*, 12, No. 2, 756-763.
25. Dondo, M. G. and El-Hawary, M. E. (1996) Application of Fuzzy Logic to Electricity Pricing in a Deregulated Environment, *Proceedings of 1996*

Canadian Conference on Electrical and Computer Engineering, Calgary, Alta., Canada, 1, 388-391.

26. Fuzzy Set Applications in Power Systems, 1997, Tutorial, *IEEE Power Industry Computer Applications Conference*, Columbus, OH.

27. Dhar, S.B. (1979) Power System Long-Range Decision Analysis under Fuzzy Environment, *IEEE Transactions on PAS*, PAS-98, No. 2, 585-596.

28. Su, C.C. and Hsu,Y.Y. (1991) Fuzzy Dynamic Programming: An Application to Unit Commitment, *IEEE Transactions on Power Systems*, 6, No. 3, 1231-1237.

29. Guan, X., Luh, P. and Prasannan, B. (1996) Power System Scheduling with Fuzzy Reserve Requirements *IEEE Transactions on Power Systems*, 11, No. 2, 864-869.

30. Miranda, V. and Saraiva, J.T. (1992) Fuzzy Modeling of Power System Optimal Power Flow, *IEEE Transactions on Power Systems*, 7, No. 2, 843-849.

31. Ramesh, V.C. and Li, X. (1997) A Fuzzy Multiobjective Approach to Contingency Constrained OPF, *IEEE Transaction on Power Systems*, 12, No. 3, 1348-1354.

32. Terano, T., Asaia, K. and Sugeno, M. (1994) Applied Fuzzy Systems, AP Professional, Cambridge, MA.

33. Umbers, I.G. and King, P.J. (1980) An Analysis of Human Decision-Making in Cement Kiln Control and the Implications for Automation, *Int. Journal of Man-Mach. Stud.*, 12, 11-23.

34. Brubakar, D. (1992) Fuzzy Environmental Control, http://www-gi.cs.cmu.e...oc/notes/brub/brub5.txt.

35. Shirmohammadi, D., Filho, X.V., Gorenstin, B. and Pereira, M.V. (1996) Some Fundamental Technical Concepts About Cost Based Transmission Pricing, *IEEE Transactions on Power Systems*, 11, No. 2, 1002-1008.

36. Yu, C.W. and David, A.K. (1997) Pricing Transmission Services in Context of Industry Deregulation, *IEEE Transactions on Power Systems*, 12, No. 1, 503-509.

37. Wu, F.F. and Varaiya, P. (1995) Coordinated Multilateral Trades for Electric Power Networks: Theory and Implementation, *UCEI Power Working Paper*, PWP-31.

38. David, A.K. (1991) An Expert System with Fuzzy Sets for Optimal Long-Range Operation Planning, *IEEE Transactions on Power Systems*, 6, 59-65.

39. Souflis, J.L., Machias, A.V. and Papadias, B.C. (1989) An Application of Fuzzy Concepts for Transient Stability Evaluation, *IEEE Transactions on Power Systems*, 4, No. 3, 1003-1009.

40. Torres, G.L., Da Siliva, L.E.B and Mukhedkar, D. (1992) A Fuzzy Knowledge Based System for Bus Load Forecasting, *IEEE International Conference on Fuzzy Systems*, San Diago, CA, 1211-1218.

41. Lin, C.E., Huang, C.L. and Lee, S.Y. (1992) An Optimal Generator Maintenance Approach Using Fuzzy Dynamic Programming, *IEEE/PES Summer Meeting*, 92 SM 401-0 PWRS, Seattle.

Chapter 8

LAGRANGIAN RELAXATION APPLICATIONS TO ELECTRIC POWER OPERATIONS AND PLANNING PROBLEMS

A.J. Conejo, J.M. Arroyo
Univ. De Castilla – La Mancha
Ciudad Real, Spain

N. Jiménez Redondo
Univ. De Málaga
Málaga, Spain

F.J. Prieto
Univ. Carlos Iii
Madrid, Spain

8.1 Introduction

This chapter presents the basic theory of both the Lagrangian Relaxation and the Augmented Lagrangian decomposition procedures. The presentation focuses mainly on algorithmic issues. To illustrate how these decomposition procedures work, three practical applications are analyzed: (i) unit commitment and short-term hydro-thermal coordination, (ii) decentralized optimal power flow, and (iii) long-term hydro-thermal coordination. Detailed bibliographical references are provided at the end of this chapter.

8.2 Lagrangian Relaxation

The Lagrangian Relaxation decomposition procedure is explained below.

8.2.1 INTRODUCTION

To advantageously apply Lagrangian Relaxation (LR) to a mathematical programming problem, the problem should have the structure below

(PP) $$\begin{aligned} \text{minimize}_x \quad & f(x) \\ \text{subject to} \quad & a(x) = 0 \\ & b(x) \leq 0 \\ & c(x) = 0 \\ & d(x) \leq 0, \end{aligned}$$ (1)

where $f(x): \Re^n \to \Re$, $a(x): \Re^n \to \Re^{\bar{a}}$, $b(x): \Re^n \to \Re^{\bar{b}}$, $c(x): \Re^n \to \Re^{\bar{c}}$, $d(x): \Re^n \to \Re^{\bar{d}}$, and $\bar{a}$, $\bar{b}$, $\bar{c}$ and $\bar{d}$ are scalars.

Constraints $c(x) = 0$, and $d(x) \leq 0$ are complicating constraints, i.e. constraints that if relaxed, problem (PP) becomes drastically simplified.

The Lagrangian function (LF) is defined as [Bazaraa93, Luenberger89]
(LF) $$L(x,\lambda,\mu) = f(x) + \lambda^T c(x) + \mu^T d(x), \qquad (2)$$
where λ and μ are Lagrange multiplier vectors.

Under local convexity assumptions ($\nabla_x^2 L(x^*, \lambda^*) > 0$) the dual function (DF) is defined as
(DF)
$$\phi(\lambda,\mu) = \text{minimum}_x \quad L(\lambda,\mu,x) \qquad (3)$$
$$\text{subject to} \quad a(x) = 0$$
$$b(x) \leq 0.$$

The dual function is concave and in general non-differentiable [Luenberger89]. This is a fundamental fact in the algorithms stated below.

The dual problem is then defined as
(DP)
$$\text{maximize}_{\lambda,\mu} \quad \phi(\lambda,\mu) \qquad (4)$$
$$\text{subject to} \quad \mu \geq 0.$$

The LR decomposition procedure is attractive if, for fixed values of λ and μ, problem (DF) is easily solved. That is, if the dual function is easily evaluated for given values $\tilde{\lambda}$ and $\tilde{\mu}$ of multiplier vectors λ and μ respectively.

The problem to be solved to evaluate the dual function for given $\tilde{\lambda}$ and $\tilde{\mu}$ is the so-called relaxed primal problem (RPP), i.e.
(RPP)
$$\text{minimize}_x \quad L(\tilde{\lambda},\tilde{\mu},x) \qquad (5)$$
$$\text{subject to} \quad a(x) = 0$$
$$b(x) \leq 0.$$

The above problem typically decomposes into subproblems. This decomposition facilitates its solution, and normally allows physical and economical interpretations. That is
(DPP)
$$\text{minimize}_{x_i, \forall i=1,\ldots,n} \quad \sum_{i=1}^{n} L_i(\tilde{\lambda},\tilde{\mu},x_i) \qquad (6)$$
$$\text{subject to} \quad a_i(x_i) = 0, \quad i = 1,\ldots,n$$
$$b_i(x_i) \leq 0, \quad i = 1,\ldots,n.$$

The above problem is called the decomposed primal problem (DPP). The resulting subproblems can be solved in parallel.

Under local convexity assumptions, the local duality theorem says that
$$f(x^*) = \phi(\lambda^*, \mu^*), \qquad (7)$$
where x^* is the minimizer for the primal problem and (λ^*, μ^*) the maximizer for the dual problem.

In the non-convex case, given a feasible solution for the primal problem, x, and a feasible solution for the dual problem, (λ, μ), the weak duality theorem says that
$$f(x) \geq \phi(\lambda, \mu). \qquad (8)$$

In the convex case, the solution of the dual problem provides the solution of the primal problem. In the non-convex case the objective function value at the optimal solution (maximizer) of the dual problem provides a lower bound to the objective function value at the optimal solution (minimizer) of the primal problem. Most electric power mathematical programming problems are non-convex.

The difference between the objective function value of the primal problem for the minimizer and the objective function value of the dual problem at the maximizer is called the duality gap. It is usually the case that the per unit duality gap decreases with the size of the primal problem [16, 6, 15].

Once the solution of the dual problem is achieved, its associated primal problem solution could be non-feasible and therefore feasibility procedures are required.

The solution of the dual problem is called *Phase 1* of the LR procedure and once known the dual optimum, the procedure to find a primal feasible near-optimal solution is called *Phase 2* of the LR procedure.

8.2.2 PHASE 1

The algorithm to solve the dual problem proceeds as follows.
- Step 0. Initialization.
  Set $v = 1$. Initialize dual variables $\lambda^{(v)} = \lambda^0$ and $\mu^{(v)} = \mu^0$. Set $\underline{\phi}^{(v-1)} = -\infty$.
- Step 1. Solution of the relaxed primal problem.
  Solve the relaxed primal problem and get the minimizer $x^{(v)}$ and the objective function value at the minimizer $\phi^{(v)}$.

Update the lower bound for the objective function of the primal problem: $\underline{\phi}^{(v)} \leftarrow \phi^{(v)}$ if $\phi^{(v)} > \underline{\phi}^{(v-1)}$.
- Step 2. Multiplier updating.
  Update multipliers using any of the procedures stated below.
- Step 3. Convergence checking.

  If $\dfrac{\|\lambda^{(v+1)}-\lambda^{(v)}\|}{\|\lambda^{(v+1)}\|+1} \leq \varepsilon$ and $\dfrac{\|\mu^{(v+1)}-\mu^{(v)}\|}{\|\mu^{(v+1)}\|+1} \leq \varepsilon$, and/or the stopping criterion of Section 1.4.5 is met, the ε-optimal solution is $x^* = x^{(v)}$, stop. Otherwise set $v \leftarrow v+1$, and go to step 1.

8.2.3 PHASE 2

Phase 2 slightly modifies multiplier values obtained at the end of *Phase 1* to achieve primal feasibility. Subgradient procedures are generally used [42]. They are effective to reach feasibility in a few iterations without altering significantly the objective function value of the dual problem at its maximizer.

8.2.4 MULTIPLIER UPDATING PROCEDURES

In this section several multiplier updating procedures are explained and compared. In the following, for the sake of clarity, multiplier vectors λ and μ are renamed as $\theta = \text{column}(\lambda, \mu)$.

The column vector of constraint mismatches at iteration v constitutes a subgradient of the dual function [3], i.e.

$$s^{(v)} = \text{column}\left[c\left(x^{(v)}\right), d\left(x^{(v)}\right)\right] \tag{9}$$

is a subgradient vector for the dual function which is used below.

8.2.4.1 Subgradient Method (SG)
The multiplier vector is updated as [32]

$$\theta^{(v+1)} = \theta^{(v)} + k^{(v)} \frac{s^{(v)}}{\left|s^{(v)}\right|} \tag{10}$$

where

$$\lim_{v \to \infty} k^{(v)} \to 0 \quad \text{and} \quad \sum_{v=1}^{\infty} k^{(v)} \to \infty. \tag{11}$$

A typical selection of $k^{(v)}$ which meets the above requirements is

$$k^{(v)} = \frac{1}{a+bv}, \qquad (12)$$

where a and b are scalar constants.

The subgradient method is simple to implement and its computational burden is small. However, it progresses slowly to the optimum in an oscillating fashion. This is a consequence of the non-differentiability of the dual function. Furthermore, the oscillating behavior makes it very difficult to devise an appropriate stopping criterion. It is typically stopped after a pre-specified number of iterations.

8.2.4.2. Cutting Plane Method (CP)

The updated multiplier vector is obtained by solving the linear programming problem below

(RDP1) $\qquad \text{maximize}_{z,\theta \in C} \quad z \qquad (13)$

$\qquad \text{subject to} \quad z \leq \phi^{(k)} + s^{(k)T}(\theta - \theta^{(k)}), \quad k = 1,\ldots,v,$

where C is a convex and compact set. It is made up of the ranges of variations of the multipliers, i.e. $C = \{\theta, \underline{\theta} \leq \theta \leq \overline{\theta}\}$. It should be noted that the above constraints represent half-spaces (hyperplanes) on the multiplier space. Values $\underline{\theta}$ and $\overline{\theta}$ are, in general, easily obtained from the physical or economical properties of the system which is modeled.

It should also be noted that the number of constraints of the above problem grows with the number of iterations.

The above problem is a relaxed dual problem (RDP) which gets closer to the actual dual problem as the number of iterations grows.

The CP method achieves a dual optimum by reconstructing the dual function. It reconstructs the region of interest and regions of no interest. This reconstruction is computationally expensive and therefore the CP method computational burden is high.

This algorithm is typically stopped when the multiplier vector difference between two consecutive iterations is below a pre-specified threshold.

8.2.4.3 Bundle Method (BD)

The updated multiplier vector is obtained solving the relaxed dual quadratic programming problem below

$\qquad \text{maximize}_{z,\theta \in C} \quad z - \alpha^{(k)} |\theta - \Theta^{(k)}|^2 \qquad (14)$

$\qquad \text{subject to} \quad z \leq \phi^{(k)} + s^{(k)T}(\theta - \theta^{(k)}), \quad k = 1,\ldots,v,$

where α is a penalty parameter and Θ, "the center of gravity", is a vector of multipliers centered in the feasibility region so that oscillations are avoided [18, 30].

It should be noted that the number of constraints of the problem above grows with the number of iterations.

The BD method is a CP method in which the ascent procedure is constrained by an objective function penalty. The target is to center the CP method in the region of interest. However, in order to do this, it is necessary to carefully tune-up the penalty and other parameters. This tune-up is problem dependent and hard to achieve.

This algorithm is typically stopped when the multiplier vector difference between two consecutive iterations is below a pre-specified threshold. More sophisticated stopping criteria are possible.

8.2.4.4 Dynamically Constrained Cutting Plane Method (DC-CP)
The updated multiplier vector is obtained solving the relaxed dual linear programming problem below

(RDP2) $\quad \text{maximize}_{z, \theta \in C^{(v)}} \quad z \hfill (15)$

$\quad \text{subject to} \quad z \leq \phi^{(k)} + s^{(k)T}\left(\theta - \theta^{(k)}\right), \quad k = 1, \ldots, n; \quad n \leq \bar{n},$

where $\bar{n}$ is the maximum number of constraints considered when solving the problem above, and $C^{(v)}$ is the dynamically updated set defining the feasibility region for the multipliers [20].

When the number of iterations is larger than the specified maximum number of constraints, the excess constraints are eliminated as stated below.

At iteration v the difference between every hyperplane evaluated at the current multiplier vector and the actual value of the objective function for the current multiplier vector (residual) is computed as

$$\varepsilon_i = \phi^{(i)} + s^{(i)T}\left(\theta^{(v)} - \theta^{(i)}\right) - \phi^{(v)} \quad \forall i = 1, \ldots, n. \hfill (16)$$

As soon as n is larger than $\bar{n}$, the "most distant" hyperplanes are not considered, so that the number of hyperplanes is kept constant and equal to $\bar{n}$. It should be noted that the residual ε_i is always positive because the cutting plane reconstruction of the dual function overestimates the actual dual function. This technique to limit the number of hyperplanes considered has proved to be computationally effective.

The dynamic updating of the set $C^{(v)}$, the feasibility region of the multipliers, is performed as stated below. Let $\theta_i^{(v)}$ be the i component of the multiplier vector at iteration v.

If $\theta_i^{(v)} = \overline{\theta}_i^{(v)}$ then   else if $\theta_i^{(v)} = \underline{\theta}_i^{(v)}$ then   (17)
$\overline{\theta}_i^{(v+1)} = \overline{\theta}_i^{(v)}(1+a)$ and $\overline{\theta}_i^{(v+1)} = \overline{\theta}_i^{(v)}(1+c)$ and
$\underline{\theta}_i^{(v+1)} = \underline{\theta}_i^{(v)}(1-b)$, $\underline{\theta}_i^{(v+1)} = \underline{\theta}_i^{(v)}(1-d)$.

Overlining indicates upper bound and underlining stands for lower bound. The parameters a, b, c and d allow to enlarge and shrink the feasibility region of the multiplier vector, i.e. the convex compact set C. This is efficiently accomplished because the above updating procedure is simple. Typical values are $a = c = 2$ and $b = d = 0.8$.

The DC-CP method is a CP method in which the ascent procedure is dynamically constrained by enlarging and shrinking the feasibility region on a coordinate basis. This is possible because the feasibility region is simple: bounds on every multiplier. Through this enlarging/shrinking procedure it is possible to center the algorithm in the area of interest which results in high efficiency. The enlarging/shrinking procedure is not problem dependent and involves straightforward heuristics (see the simple updating procedure above). This method is actually a trust region procedure with dynamic updating of the trust region.

This algorithm is typically stopped when the multiplier vector difference between two consecutive iterations is below a pre-specified threshold. A small enough difference between an upper bound and a lower bound of the dual optimum is also an appropriate stopping criterion.

8.2.4.5 Stopping Criterion
When using the CP or the DC-CP method, at every iteration, the objective function value of the relaxed dual problem constitutes an upper bound of the optimal dual objective function value. This is so because the piecewise linear reconstruction of the dual function overestimates the actual dual function. On the other hand, the objective function value of the dual problem (evaluated through the relaxed primal problem) provides at every iteration a lower bound of the optimal dual objective function value. This can be mathematically stated as follows

$$z^{(v)} \geq \phi^* \geq \phi^{(v)} \qquad (18)$$

where $z^{(v)}$ is the objective function value of the relaxed dual problem at iteration v, ϕ^* is the optimal dual objective function value, and $\phi^{(v)}$ is the objective function value of the dual problem at iteration v.
The size of the per unit gap

$$g^{(v)} = \left(z^{(v)} - \phi^{(v)}\right) / \left(\phi^{(v)} + 1\right) \tag{19}$$

is an appropriate objective function value criterion to stop the search for the dual optimum.

8.2.5 EXAMPLE

To clarify how the LR decomposition procedure works, a simple example is solved in this subsection.

Consider the problem

$$\begin{aligned}
\text{minimize}_{x,y} \quad & x^2 + y^2 = f(x,y) \\
\text{subject to} \quad & x + y - 10 = 0 \\
& x \geq 0, \quad y \geq 0,
\end{aligned}$$

whose solution is $x^* = y^* = 5$, $f(x^*, y^*) = 50$. The Lagrange multiplier associated to the equality constraint has an optimal value $\lambda^* = -10$.

The Lagrangian function is

$$L(x, y, \lambda) = x^2 + y^2 + \lambda(x + y - 10).$$

The solution of this problem by LR follows.
- Step 0. Initialization.

    Set $\lambda = \lambda^0$.
- Step 1. Solution of the relaxed primal problem.

    This problem decomposes into the two subproblems below (-5λ is arbitrarily assigned to each subproblem)

$$\begin{array}{ll}
\text{minimize}_x \quad x^2 + \lambda x - 5\lambda & \qquad \text{minimize}_y \quad y^2 + \lambda y - 5\lambda \\
\text{subject to} \quad x \geq 0, & \qquad \text{subject to} \quad y \geq 0,
\end{array}$$

   whose solutions are respectively x^c and y^c.
- Step 2. Multiplier updating.

    Use a subgradient procedure with proportionality constant equal to δ.

$$\lambda \leftarrow \lambda + \delta(x^c + y^c - 10).$$

- Step 3. Convergence checking.

If multiplier λ does not change sufficiently, stop; the optimal solution is $x^* = x^c, y^* = y^c$. Otherwise the procedure continues in Step 1.

Considering $\delta = 1$ and an initial multiplier value $\lambda = 1$, the algorithm proceeds as in Table 1 below.

TABLE 1. Example: Evolution of the Lagrangian Relaxation algorithm

Iteration #	λ	x	y	f(x,y)	L(x,y,λ)
1	1	0.0	0.0	00.0	-10.0
2	-9	4.5	4.5	40.5	49.5
3	-10	5.0	5.0	50.0	50.0

8.3 Augmented Lagrangian Decomposition

The Augmented Lagrangian decomposition procedure is explained below.

The Augmented Lagrangian (AL) function of problem (1) has the form

(AL)
$$A(x,\lambda,\mu,\alpha,\beta) = f(x) + \lambda^T c(x) + \mu^T \tilde{d}(x,z) + \frac{1}{2}\alpha\|c(x)\|^2 + \frac{1}{2}\beta\|\tilde{d}(x,z)\|^2. \quad (20)$$

Penalty parameters α and β are large enough scalars to ensure local convexity, the component i of function $\tilde{d}(x,z)$ is defined as $\tilde{d}_i(x,z) = d_i(x) + z_i^2$, and z_i's $(i = 1,\ldots,\bar{d})$ are additional variables to transform inequality constraints into equality constraints.

It should be noted that the quadratic terms confer the Augmented Lagrangian function good local convexity properties.

The decomposition based on the Augmented Lagrangian is basically similar to the Lagrangian Relaxation decomposition procedure stated in the previous section. The basic difference is that quadratic terms in the Augmented Lagrangian make the relaxed primal problem non-decomposable. To make it decomposable two procedures are mainly used.

The first procedure linearizes the quadratic terms of the Augmented Lagrangian and fixes the minimum number of variables to the values of the previous iteration to achieve separability [8]. The second procedure directly fixes in the Augmented Lagrangian the minimum number of variables to the values of the previous iteration to achieve separability.

Being the Augmented Lagrangian locally convex, a gradient (not a subgradient) of the dual function is available at every iteration, i.e.

$$g^{(v)} = \text{column}\left[c(x^{(v)}), \tilde{d}(x^{(v)}, z^{(v)})\right] \quad (21)$$

is a gradient of the dual function.

An appropriate rule to update multipliers λ is therefore [22]
$$\lambda^{(v+1)} = \lambda^{(v)} + \alpha c(x^{(v)}). \qquad (22)$$

Multipliers μ can be updated as
$$\mu^{(v+1)} = \mu^{(v)} + \beta \tilde{d}(x^{(v)}, z^{(v)}). \qquad (23)$$
It should be noted that variables z_i's can be treated implicitly, as stated in [5].

Further details on multiplier updating procedures can be found in [22] and [5].

8.3.1 EXAMPLE

To clarify how the Augmented Lagrangian decomposition procedure works, a simple example is solved in this subsection. The problem considered is the one stated in Subsection 1.5.

The Augmented Lagrangian has the form
$$A(x, y, \lambda, \rho) = x^2 + y^2 + \lambda(x + y - 10) + \rho(x + y - 10)^2.$$
Being $\bar{x}$ and $\bar{y}$ estimates for x and y respectively, the Augmented Lagrangian can be written as
$$A(x, y, \lambda, \rho) = x^2 + \lambda x - 5\lambda + \frac{1}{2}\rho(x + \bar{y} - 10)^2 + y^2 + \lambda y + -5\lambda + \frac{1}{2}\rho(\bar{x} + y - 10)^2.$$
The solution of this problem through the Augmented Lagrangian procedure follows.
- Step 0. Initialization.
  Set $\lambda = \lambda^0$, $\bar{x} = x^0$ and $\bar{y} = y^0$.
- Step 1. Solution of the relaxed primal problem.
  This problem decomposes into the two subproblems below

$$\text{minimize}_x \quad x^2 + \lambda x - 5\lambda + \frac{1}{2}\rho(x + \bar{y} - 10)^2$$
$$\text{subject to} \quad x \geq 0,$$

$$\text{minimize}_y \quad y^2 + \lambda y - 5\lambda + \frac{1}{2}\rho(\bar{x} + y - 10)^2$$
$$\text{subject to} \quad y \geq 0,$$

whose solutions are respectively x^c and y^c.
- Step 2. Multiplier updating.
  Use a gradient procedure with a proportionality constant equal to δ.

$$\lambda \leftarrow \lambda + \delta(x^c + y^c - 10).$$

- Step 3. Convergence checking.

  If multiplier λ does not change sufficiently, stop; the optimal solution is $x^* = x^c$, $y^* = y^c$. Otherwise, update the estimates of x and y, for instance, imposing that the relaxed constraint is met, which results in $\overline{x} = \frac{1}{2}(x^c - y^c + 10)$, $\overline{y} = \frac{1}{2}(x^c - y^c + 10)$; the procedure continues in Step 1.

Considering $\rho = 10$, $\delta = 5$, an initial multiplier value $\lambda = 1$, and initial variable estimates $\overline{x} = 2$, $\overline{y} = 2$, the algorithm proceeds as shown in Table 2 below.

TABLE 2. Example: Evolution of the Augmented Lagrangian algorithm

Iteration #	x	y	λ	f(x,y)	A(x,y,λ,ρ)
1	6.58	6.58	1.00	86.68	109.92
2	2.76	2.76	16.83	15.28	-10.00
3	4.63	4.63	-5.53	42.82	48.33
4	4.94	4.94	-9.26	48.76	49.95
5	4.99	4.99	-9.88	49.79	49.99
6	5.00	5.00	-9.98	50.00	50.00
7	5.00	5.00	-10.00	50.00	50.00

8.4 Short-Term Hydro-Thermal Coordination and Unit Commitment

8.3.1 INTRODUCTION

Short-term hydro-thermal coordination (STHTC) determines the start-up and shut-down of thermal plants, as well as the power output of hydro and thermal plants to meet customer demand with an appropriate level of security and so that total operating costs are minimized. Because of the high cost associated to the start-up of thermal plants, selecting the plants to meet customer demand in the most economical manner can save large amounts of money. If the electric energy system under consideration does not include hydroelectric plants, the above problem is called unit commitment.

Mathematically, the STHTC problem can be formulated as a mixed-integer nonlinear optimization problem. For realistic size electric energy systems it is also a large-scale problem. Solving this large-scale nonlinear and combinatorial optimization problem is not an easy task. Lagrangian Relaxation techniques are the most suitable techniques to solve this kind of problems [29, 28, 6, 42, 41, 27, 33, 39, 30, 21, 20]. Dynamic programming techniques require discretization of continuous variables and drastic simplifying assumptions to make the problem computationally tractable [19]. Mixed-integer linear programming techniques not only linearize the problem but also make important simplifications in order to be able to solve such a large-scale problem [12, 7, 23, 24, 25, 26].

When using LR techniques to solve the STHTC problem, the resulting relaxed primal problem can be naturally decomposed into one subproblem per thermal plant and one subproblem per hydro system. Therefore, by using LR techniques, the solution of the STHTC problem (large-scale and complex optimization problem) is accomplished by the solution of many small size and structurally homogeneous subproblems.

This decomposition property allows a very precise modeling of each generating plant as well as the possibility to apply to each subproblem the most suitable optimization technique to its structure. It also allows the natural application of parallel computing with the corresponding advantages regarding CPU time.

Apart of all these advantages derived from the decomposition property of the relaxed primal problem, the application of LR techniques to solve the STHTC problem presents another important advantage: the dual problem variables (the Lagrange multipliers) have an economical meaning which can be very helpful in the framework of deregulated electric energy markets, and also in the traditional framework of centralized systems.

8.4.2 PROBLEM FORMULATION AND LR SOLUTION PROCEDURE

The STHTC problem can be formulated as a nonlinear and combinatorial optimization problem in which total operating costs are minimized subject to meet constraints modeling the technical limitations of thermal and hydro plants, and to meet load constraints. Load constraints include electric energy customer demand constraints plus spinning reserve constraints. Spinning reserve constraints ensure an appropriate level of security. This problem will be referred to as the primal problem (PP) and is formulated as follows

$$\begin{aligned}
& \underset{x=(x_i,x_j)}{\text{minimize}} \quad f(x) = \sum_{i=1}^{n} f_i(x_i) \\
& \text{subject to} \quad s_i(x_i) \leq 0 \qquad i = 1,\ldots,I \\
& \qquad\qquad\quad s_j(x_j) \leq 0 \qquad j = 1,\ldots,J \\
& \qquad\qquad\quad \sum_{i=1}^{I} h_i(x_i) + \sum_{j=1}^{J} h_j(x_j) = H \\
& \qquad\qquad\quad \sum_{i=1}^{I} g_i(x_i) + \sum_{j=1}^{J} g_j(x_j) \leq G,
\end{aligned} \qquad (24)$$

where x_i is the vector of variables associated to thermal plant i, x_j is the vector of variables associated to hydro system j, I is the number of thermal plants, and J is the number of hydro systems. H, $h_i(x_i)$, $h_j(x_j)$, G, $g_i(x_i)$ and $g_j(x_j)$ are vectors of dimension equal to the number of time periods in the planning horizon. A typical hydro system is shown in figure 1.

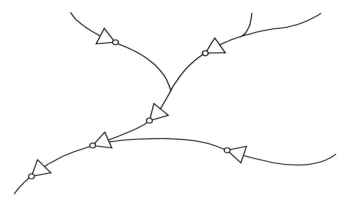

Figure 1. Typical river system

In the formulation above, the objective function represents the total operating cost (the cost of hydro power production is negligible compared to the cost of thermal power production). The first set of constraints expresses thermal plant constraints, the second one represents hydro system constraints, the third one expresses customer demand constraints, and the fourth one represents spinning reserve constraints. It should be noted that time is embedded in the above formulation.

Load constraints are the complicating or global constraints of the above problem. They couple together decisions related to thermal and hydro plants. Due to the existence of these constraints the above problem cannot be decomposed and cannot be easily solved. Load constraints include equality constraints (demand constraints) and inequality constraints (spinning reserve constraints).

By applying LR techniques, load constraints are incorporated into the objective function to form the relaxed primal problem. The vector of multipliers associated to the vector of demand constraints is called λ and the vector of multipliers associated to the spinning reserve constraints is called μ.

The Lagrangian function is defined as

$$L(x,\lambda,\mu) = \sum_{i=1}^{I} f_i(x_i) + \lambda^T \left(H - \sum_{i=1}^{I} h_i(x_i) - \sum_{j=1}^{J} h_j(x_j) \right) + \mu^T \left(G - \sum_{i=1}^{I} g_i(x_i) - \sum_{j=1}^{J} g_j(x_j) \right), \quad (25)$$

and the dual function is the solution of the problem

$$\phi(\lambda,\mu) = \text{minimize}_{(x_i, x_j)} \quad L(\lambda,\mu,x_i,x_j) \qquad (26)$$
$$\text{subject to} \quad s_i(x_i) \le 0$$
$$s_j(x_j) \le 0,$$

which can be expressed as

$$\phi(\lambda,\mu) = \lambda^T H + \mu^T G + d(\lambda,\mu) \qquad (27)$$

where $d(\lambda,\mu)$ is the solution of the following optimization problem

$$\text{minimize}_{(x_i, x_j)} \sum_{i=1}^{I}\left(f_i(x_i) - \lambda^T h_i(x_i) - \mu^T g_i(x_i)\right) - \qquad (28)$$
$$\sum_{j=1}^{J}\left(\lambda^T h_j(x_j) + \mu^T g_j(x_j)\right)$$
$$\text{subject to} \quad s_i(x_i) \le 0 \quad i=1,\ldots,I$$
$$s_j(x_j) \le 0 \quad j=1,\ldots,J.$$

The above problem can be naturally decomposed into one subproblem per thermal plant i and one subproblem per hydro system j. For fixed values of λ and μ, problem (26) is called the relaxed primal problem, and problem (28) is called the decomposed primal problem.

The subproblem associated to thermal plant i is

$$\text{minimize}_{x_i} \quad f_i(x_i) - \lambda^T h_i(x_i) - \mu^T g_i(x_i) \qquad (29)$$
$$\text{subject to} \quad s_i(x_i) \le 0,$$

and the subproblem associated to hydro system j is

$$\text{maximize}_{x_j} \quad \lambda^T h_j(x_j) + \mu^T g_j(x_j) \qquad (30)$$
$$\text{subject to} \quad s_j(x_j) \le 0.$$

LR techniques are based on the solution of the following dual problem

$$\text{maximize}_{(\lambda,\mu)} \quad \phi(\lambda,\mu) \qquad (31)$$
$$\text{subject to} \quad \mu \ge 0.$$

Due to the existence of integer variables (e.g. thermal plant unit commitment variables) in the formulation of the primal problem, the STHTC is a non-convex problem. Therefore, the optimal solution of the dual problem is not the optimal solution of the primal problem but it is a lower bound. Nevertheless, as the size of the problem increases the per unit duality gap (see Section 1.1) decreases [16, 6, 15], and therefore the optimal solution of the dual problem becomes closer to the optimal solution of the primal problem. Once the optimal solution of the dual problem is found, heuristic procedures can be easily applied to derive a near-optimal primal problem solution [42].

In the STHTC problem, inequality complicating constraints are closely related to the integer variables (the thermal plant unit commitment variables) and the equality constraints are closely related to the continuous variables (thermal and hydro plant power output variables). This motivates the decomposition of *Phase 2* (in Section 1.3) into two consecutive phases. In the first one, called *Phase 2A*, the solution of the dual problem (or *Phase 1*) is slightly modified to find values of the integer variables that meet inequality global constraints (spinning reserve constraints). In the second one, called *Phase 2B*, the solution of *Phase 2A* is modified by adjusting the values of the continuous variables to meet equality global constraints (demand constraints).

Therefore, the LR procedure to solve the STHTC problem consists of:
- *Phase 1*: Solution of the dual problem.
- *Phase 2A*: Search for a spinning reserve primal feasible solution.
- *Phase 2B*: Search for a load balanced primal feasible solution: multiperiod economic dispatch.

The solution of the dual problem (*Phase 1*) is the key element to solve the STHTC problem by LR. The efficiency of a STHTC algorithm relies on the efficiency of the solution of this phase.

Phase 2A is an iterative procedure in which μ multipliers are updated in those periods where spinning reserve constraints are not satisfied. In these periods, the corresponding μ multipliers are increased proportionally to the mismatches in spinning reserve constraints (subgradient type updating) until these constraints are met in all the time periods of the planning horizon. At the end of *Phase 2A* a primal set of feasible commitment decisions is found. This phase requires typically little CPU time to achieve a solution very close to the solution of *Phase 1*.

Phase 2B is a multiperiod economic dispatch procedure [40] in which, once unit commitment variables are set to the solution of *Phase 2A*, power output is adjusted in order to meet the demand constraints in all time periods. This is a traditional problem routinely solved by electric energy system operators.

8.4.3 DUAL PROBLEM SOLUTION: MULTIPLIER UPDATING TECHNIQUES

The procedure to solve the dual problem of the STHTC problem is described in Section 1.2. That is, at each iteration the relaxed primal problem is solved and, with the information obtained, the multiplier vector is updated. The information derived from the resolution of the relaxed primal problem is
- the value of the dual function

$$\phi(\lambda^{(v)},\mu^{(v)})=\lambda^T H+\mu^T G+d(\lambda^{(v)},\mu^{(v)}), \qquad (32)$$

being

$$d(\lambda^{(v)},\mu^{(v)}) = \sum_{i=1}^{I} \left(f_i(x_i^*) - \lambda^T h_i(x_i^*) - \mu^T g_i(x_i^*)\right) - \qquad (33)$$
$$\sum_{j=1}^{J} \left(\lambda^T h_j(x_j^*) + \mu^T g_j(x_j^*)\right),$$

where x_i^* is the vector of optimal values for the variables associated to thermal plant i, and x_j^* is the vector of optimal values for the variables associated to hydro system j obtained from the solution of the relaxed primal problem; and
- a subgradient $s^{(v)}$ of the dual function at the optimal solution of the relaxed primal problem, i.e. $x_i^* \ \forall i, \ x_j^* \ \forall j$.

A subgradient can be easily computed as the vector of mismatches in demand constraints and the vector of mismatches in spinning reserve constraints, that is

$$s^{(v)} = \text{column}\left[h^{(v)}, g^{(v)}\right], \qquad (34)$$

where

$$h^{(v)} = H - \sum_i h_i(x_i^{*(v)}) - \sum_j h_j(x_j^{*(v)}) \qquad (35)$$
$$g^{(v)} = G - \sum_i g_i(x_i^{*(v)}) - \sum_j g_j(x_j^{*(v)}).$$

Though the four methods described in Section 1.4 can be applied to update the multipliers, the most suitable methods for the STHTC problem are the bundle method [30] and the dynamically constrained cutting plane method [20].

8.4.4 ECONOMICAL MEANING OF THE MULTIPLIERS

As it has been indicated above one of the advantages of the use of LR techniques is the availability of the useful economical information provided by the variables of the dual problem, the Lagrange multipliers.

Multiplier λ at a given time represents, from the point of view of the system, the cost of producing one extra unit of electric energy (MWh), that is, the electric energy marginal cost. Equivalently, from the point of view of a generating company, multiplier λ at a given time represents an indicator of the price a plant should be paid for each MWh of energy. It also represents an indicator of the price that a generating company should bid to get its plant on line.

Analogously, multiplier µ at a given time represents the cost of keeping an incremental unit (MW) of power reserve or equivalently, an indicator of the price a generator should be paid for each MW of reserve.

This economical interpretation is useful in the traditional framework of centralized electric energy systems to elaborate electric tariffs, but also in the framework of modern deregulated electric energy markets.

In the framework of deregulated electric energy markets, the LR procedure to solve the STHTC problem can be interpreted as the actual functioning of a free market. In other words, it can be thought as a mechanism to meet customer demand with an appropriate level of security (measured in terms of the spinning reserve) by choosing the cheapest generator offers.

Hourly energy prices proposals (Lagrange multipliers) are specified by the market operator for the planning horizon. Each generator (or generating company) schedules independently its production along the planning horizon to maximize its benefit (i.e. each generator solves an optimization problem of type (29)). Analogously, each hydro system is scheduled so that its benefit is maximum (i.e. each hydro system solves a problem of type (30)). After the submissions of production proposals by all generators the demand equation is evaluated in each hour of the planning horizon. Hourly prices are updated by the market operator with any of the techniques stated above, and the previous procedure is repeated until the demand is satisfied. This mechanism constitutes a competitive energy market. Similarly a spinning reserve market can be established.

It should be noted that by applying the LR technique to solve the STHTC problem, each generator schedules its production attending only to prices of energy and prices of reserve. The interchange of information between the market operator and the generators is clear and concise. The resulting market is therefore economically efficient and transparent.

The Augmented Lagrangian decomposition technique has also been applied to the STHTC problem. Relevant information can be found in [1, 2, 34, 39].

8.5 Decentralized Optimal Power Flow

8.5.1 INTRODUCTION

With the introduction of deregulation in the electricity sector, the idea of an electric energy system strongly coordinated by one entity is being replaced by a market where independence among utilities is one of the main features. This process of deregulation affects all the problems involved in the operation and planning of an electric energy system. One of these problems is the optimal power flow (OPF). Typically, an electric energy system is divided into several areas (see figure 2) owned by different utilities. In the new competitive framework, electrical utilities are interested in solving the optimal power flow problem corresponding to its own area without considering the rest of areas owned by the other competing electrical utilities. Thus, the traditional centralized optimal power flow problem becomes a decentralized optimal power flow problem. However, to maintain economic efficiency, the optimal solution of the decentralized OPF should be identical to the optimal solution of the centralized OPF.

The Lagrangian Relaxation decomposition procedure is used to solve the OPF in a decentralized and coordinated fashion. The OPF of every area is solved by its own utility and the coordinator is just needed to interchange a small amount of information among the utilities involved. Therefore, the optimal solution (economic efficiency) is obtained and, at the same time, the independence required by the utilities is maintained. Under the coordinator point of view, the Lagrangian Relaxation decomposition procedure enables to solve the OPF in a decentralized but coordinated way. Thus, the subproblem of each area can be solved independently and, consequently, in a parallel way. Moreover, optimal energy pricing rates (Lagrange multipliers) for the energy traded through the interconnections are derived.

It should be clear that the decomposition is not oriented to improve computational efficiency but to preserve dispatching independence of each area in a multi-area electric energy system.

In what follows a DC model of the OPF problem is formulated and the application of the Lagrangian Relaxation decomposition procedure is explained in detail.

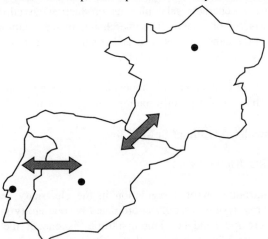

Figure 2. Multi-area electric energy system

8.4.2. PROBLEM FORMULATION

The optimal power flow problem consists of determining the power production of every generator and the flow through each transmission line of the electric energy system. This should be accomplished meeting the power balance constraints at every bus, and the limits in generation capacities and transmission line capacities in such a way that the total generation cost is minimized. Several models are used for this problem. In this work a DC model [40] is considered. This means that only real power is taken into account. Losses are incorporated through additional loads based on cosine approximations. This approximation is accurate enough for the purpose of this application. Detail information on more sophisticated formulations can be found in [40].

The mathematical formulation of the multi-area DC OPF problem has the form [9]

$$\text{minimize} \quad \sum_{i \in \{U_a \Lambda_a\}} f_i(p_i) \quad (36)$$

$$\text{subject to} \quad p_i - D_i + \sum_{k \in \Omega_i} [B_{ik}(\delta_k - \delta_i) - L_{ik}(1 - \cos(\delta_k - \delta_i))] = 0 \quad \forall i \in \{U_a \Theta_a\}$$

$$-\overline{P}_{ij} \leq B_{ij}(\delta_i - \delta_j) \leq \overline{P}_{ij} \quad \forall i \in \{U_a \Theta_a\} \; \forall j \in \{U_a \Theta_a\} i \neq j$$

$$\underline{P}_i \leq p_i \leq \overline{P}_i \quad \forall i \in \{U_a \Lambda_a\},$$

where p_i is the power produced by generator i, δ_i is the angle of bus i, $f_i(\cdot)$ is the operating cost of generator i, D_i is the demand at bus i, B_{ik} is the inverse of the reactance of line ik, $\overline{P}_{ij}$ is the maximum transmission capacity of line ij, $\underline{P}_i$ is the minimum power output of generator i, $\overline{P}_i$ is the maximum power output of generator i, Λ_a is the set of generators of area a, Θ_a is the set of buses of area a, Ω_i is the set of buses connected to bus i, and U_a is the set union over sets with subindex a.

The loss coefficient of line ij, L_{ij}, is

$$L_{ij} = \frac{R_{ij}}{R_{ij}^2 + (1/B_{ij})^2} \quad (37)$$

where R_{ij} is the resistance of line ij.

The objective function is the addition of generator operating costs. The first block of constraints enforces power balance at every bus, the second block enforces transmission capacity limits for every line, and the third block states bounds on power output for every generator.

In order to make this dispatching primal problem separable per area a fictitious bus per interconnecting line is included. The purpose of the fictitious buses is to preserve the integrity of the particular dispatching problem of every area while allowing area decomposition.

The formulation of the problem is slightly modified due to the inclusion of fictitious buses. The following equations must be included in the formulation

$$h_{ij} = 2B_{ij}(\delta_i - \delta_x) + 2B_{ij}(\delta_j - \delta_x) = 0 \quad : \alpha_{ij} \quad \forall ij \in \{U_a \Xi_a\} \quad (38)$$

where x represents a fictitious bus, α_{ij} is the multiplier associated to the power balance equation of the fictitious bus of line ij, and Ξ_a is the set of interconnecting lines of area a. The term $2B_{ij}(\delta_i - \delta_x)$ represents the power flow from i to x and, analogously, the term $2B_{ij}(\delta_j - \delta_x)$ represents the power flow from j to x.

These new equations enforce power balance at the fictitious buses. It should be noted that loads representing losses are placed at the non-fictitious buses of the interconnecting lines.

As stated below, the Lagrange multipliers associated to constraints (38) are interpreted as optimal energy trading prices at interconnecting lines. Additionally, this approach allows an independent but sequential solution of the area subproblems. The sequential solution of the area subproblems can be converted into a completely parallel solution by using an Augmented Lagrangian decomposition procedure.

It should also be mentioned that the solution provided by the decentralized formulation is identical to the solution obtained by using a centralized approach.

8.5.3 SOLUTION APPROACH

In terms of the DC OPF problem, the complicating constraints are the coupling constraints among areas (power balance at the fictitious buses). They are incorporated into the objective function through Lagrange multipliers building the Lagrangian function

$$L = \sum_{i \in \{U_a \Lambda_a\}} f_i(p_i) + \sum_{ij \in \{U_a \Xi_a\}} \alpha_{ij}(2B_{ij})[(\delta_i - \delta_x) + (\delta_j - \delta_x)] \quad (39)$$

The resulting relaxed primal problem can be formulated as follows

$$\text{minimize} \quad \sum_{i \in \{U_a \Lambda_a\}} f_i(p_i) + \sum_{ij \in \{U_a \Xi_a\}} \alpha_{ij}(2B_{ij})[(\delta_i - \delta_x) + (\delta_j - \delta_x)] \quad (40)$$

subject to $p_i - D_i + \sum_{k\in\Omega_i}[B_{ik}(\delta_k - \delta_i) - L_{ik}(1 - \cos(\delta_k - \delta_i))] = 0 \quad \forall i \in \{U_a \Theta_a\}$

$-\overline{P}_{ij} \leq B_{ij}(\delta_i - \delta_j) \leq \overline{P}_{ij} \quad \forall i \in \{U_a\Theta_a\} \forall j \in \{U_a\Theta_a\}, i \neq j$

$\underline{P}_i \leq p_i \leq \overline{P}_i \quad \forall i \in \{U_a \Lambda_a\}.$

This problem decomposes by area preserving therefore the dispatching independence of each area. It is therefore a decomposed primal problem. This is the main advantage of solving the dual problem instead of solving the primal problem (see Section 1.1).

The formulation of the independent subproblem of each area is stated as follows

minimize $\sum_{i\in\Lambda_a} f_i(p_i) + \sum_{ij\in\Xi_a} \alpha_{ij}(2B_{ij})(\delta_i - \delta_x)$ (41)

subject to $p_i - D_i + \sum_{k\in\Omega_i}[B_{ik}(\delta_k - \delta_i) - L_{ik}(1 - \cos(\delta_k - \delta_i))] = 0 \quad \forall i \in \Theta_a$

$-\overline{P}_{ij} \leq B_{ij}(\delta_i - \delta_j) \leq \overline{P}_{ij} \quad \forall i \in \Theta_a, \forall j \in \Theta_a, i \neq j$

$\underline{P}_i \leq p_i \leq \overline{P}_i \quad \forall i \in \Lambda_a.$

From the objective function of the problem above, α_{ij} can be interpreted as the buying/selling price of energy at the fictitious bus of interconnecting line ij. Therefore, the multiplier α_{ij} is the optimal energy pricing rate for the energy traded through line ij once the decomposition procedure has converged.

The consequence of considering just one fictitious bus per interconnecting line is the sequential concatenation of each area subproblem, therefore a sequential solution of area subproblems is required.

A mechanism is required to update the Lagrange multipliers so that the dual problem is maximized. The procedure for updating multipliers is the task carried out by the coordinator of the system. Several methods for updating multipliers can be used. In this case, a subgradient method has been successfully applied. Although the subgradient method presents a slow rate of convergence, in this particular application, it has shown a good performance [9].

In conclusion, the process to solve the decentralized and coordinated DC OPF problem implies a rather small exchange of information between the coordinator and the utilities. Every utility solves its own DC OPF for fixed values of interconnecting prices (Lagrange multipliers) and sends the coordinator just the mismatches in the power balance equations at the fictitious buses (subgradient vector). With this information, the coordinator returns to every utility a new value for each interconnecting price based on the current mismatches. This process is repeated until the stopping criteria are met.

8.6 Medium/Long-Term Hydrothermal Coordination Problem

In this section an application of Augmented Lagrangian techniques to the solution of a medium/long-term hydrothermal coordination problem under uncertainty is described. This is a large-scale optimization problem with a structure that allows for the decomposition of the original problem into subproblems, whose properties can be exploited by the optimization code.

8.6.1 THE PROBLEM

The goal of the medium/long-term hydrothermal coordination problem is the optimization of the operation of a hydrothermal energy generating system over a time horizon of one to four years. This problem has been extensively treated in the literature [37, 10, 38, 13].

Hydro generation has very low operation costs, but its availability is not directly controlled by the generating companies. In fact, this availability may present a very high variability along time. As a consequence, when the hydro generation is a significant percentage of total generation, it is important, and particularly so in a competitive environment, to be able to use a planning tool to decide on the best usage of this resource in the short term, taking into account future uncertainties. This same approach can be applied to other generating resources whose procurement decisions are taken over the medium term, but their usage is decided in the short term, such as fuel procurement, energy exchange contracts, etc.

The uncertainty in the availability of the water in the reservoir systems is an important aspect to be modeled. If the variation in water availability from one year to another is large, there is a significant amount of risk involved when planning on the basis of some average set of values. In a competitive environment it is important to consider the uncertainty in the model in order to both exploit the competitive advantages of this resource, and take into account a certain amount of risk management.

The model presented emphasizes the modeling of the uncertainty in the exogenous water inflows, to the expense of other aspects in the model. For example, the thermal system is represented through a savings function, obtained from a merit ordering of the thermal units. In this way, a model is derived where the constraints have a network structure. This structure can be exploited to improve the computational efficiency of the solution process, and this is important to allow for a more detailed model of the uncertainty.

8.6.2 THE TREATMENT OF UNCERTAINTY

One of the main motivations to model the uncertainty within the optimization problem is the significance to the solution of considering not just the most likely behavior of the system, but also to model explicitly the possible deviations from this average behavior.

To achieve this, the model considers simultaneously a set of outcomes. The information associated with each one of the outcomes must be combined to yield a meaningful solution, that is, one that can be used in practice [35]. This combination can be made in different ways.

One approach is to use Benders decomposition techniques together with either Montecarlo methods [36] or importance sampling to build an approximation to the expected value of the variables of interest (cost, benefits, etc.), or to some higher order moments, as might be needed for risk avoidance purposes. Some illustrations of this approach are found in [31, 11].

Alternatively [43], it is possible to work with a set of scenarios that represent the different outcomes to be considered within the planning process. These scenarios are chosen either to depict accurately the distribution of possible values, or to cover extreme situations, in order to provide a measure of risk protection within the plan.

The scenarios can be generated using a predictor module for the water inflows, based, for instance, on a neural network model, and then introducing perturbations on the predicted values. The perturbations are modeled to reproduce (in an approximate manner) the variability that is present in the historical data, as well as the temporal and spatial correlations in the data.

To model the availability of information through time, the scenarios are defined to share all data in the initial periods, and then to differentiate progressively through time. The initial period has a single scenario, and for the remaining periods the scenarios present a tree structure, where every few periods there is a differentiation between groups of scenarios by introducing a reduced number (two or three) of different outcomes for the uncertainty in that time period.

The solutions corresponding to the different scenarios are combined according to an information availability criterion. The nonanticipativity condition requires all solutions corresponding to scenarios that have the same values of the uncertain variables up to a given time (all scenarios corresponding to the same node in the scenario tree for that period), to have the same solutions up to that time.

This condition can be imposed in different ways. In order to preserve as much as possible the network structure in the constraints, the constraints below are included

$$x^s = x^{s'}, \tag{42}$$

where x^s represents the system variables at scenario s. Superscript s' indicates a different scenario from scenario s, but satisfying the nonanticipativity condition.

8.6.3. THE MODEL

After incorporating the elements described in the preceding paragraphs, the resulting model has the form

$$\text{maximize}_x \quad \sum_s w_s \sum_t f_t(x^s) \tag{43}$$
$$\text{subject to} \quad Ax^s = b^s$$
$$l^s \le x^s \le u^s$$
$$x^s = x^{s'}.$$

This model includes an objective function defined in terms of the expected thermal savings that incorporates information on the availability of the units. This function has been approximated through a polynomial, $f_i(\cdot)$; it is nonconvex, but separable for different time periods and different scenarios. The parameter w_s is the weight assigned to scenario s.

The linear constraints corresponding to the water balance in the reservoirs are separable for different scenarios, and have network structure. These constraints are expressed through matrix A and a right-hand side vector b^s. They are also separable by river system. Given the values of the state variables, the reservoir levels, they would be separable by time period. Constant vectors l^s and u^s represent respectively upper and lower bounds on variables x^s.

The complicating constraints added to take into account the relationship between the different scenarios (42) only affect the state variables, that is, the reservoir levels. These complicating constraints destroy the network structure of the constraints, but if they are removed from the problem, a model separable by scenario is obtained.

As a consequence, a nonlinear and nonconvex model is obtained. This model has linear constraints such that if some of them are removed, it recovers a network structure, it becomes separable by scenario and river system and, by fixing the values of some of the variables, it also becomes separable by time period.

The problem, for a generating system of reasonable size (50 thermal and 50 hydro units) in which the uncertainty is modeled with a sufficient amount of detail for a high variability situation (200-1000 scenarios), can reach a very large size. For example, if for the preceding system 15 time periods are considered (4 weekly plus 11 monthly periods) the resulting problem has in the order of one million variables and half a million constraints.

Given the size of the full problem, and the loss of the network structure, it is reasonable to apply some decomposition procedure that is able to solve the problems corresponding to each scenario separately. These problems are much smaller in size, having for the preceding example 2500 variables and 750 constraints, and preserving the network structure in the constraints. Another advantage of this decomposition approach is the possibility of using parallel or distributed computation to speed up the solution process.

8.6.4. DECOMPOSITION PROCEDURES

Several decomposition procedures have been proposed in the literature to solve problems with these characteristics. In particular, Benders decomposition has been used extensively, and Lagrangian Relaxation techniques have also been applied in many cases. Lately, Augmented Lagrangian techniques have been used [43, 13].

If the medium-term hydrothermal coordination problem is implemented using this last technique, it presents a very low overhead for the updating of the data between iterations. This is particularly so if subgradient based updating procedures are used, and this simplifies the parallel implementation of the code and allows for a higher speedup. At the same time, the use of an Augmented Lagrangian provides a faster and smoother convergence to the solution.

Given the structure of the problem presented in (43), a direct application of the Augmented Lagrangian techniques leads to the removal into the objective function of the constraints included in (42). As a consequence, the resulting problem has the form

$$\text{maximize}_x \quad \sum_s w_s \sum_t f_t(x^s) - \sum_s \lambda^s (x^s - x^{s'}) + \frac{\rho}{2} \sum_s \left\| x^s - x^{s'} \right\|^2 \quad (44)$$

$$\text{subject to} \quad Ax^s = b^s$$

$$l^s \le x^s \le u^s.$$

Once the problem is in this form, and ignoring for the moment the quadratic term, it is possible to separate it by scenario, and a significant size reduction can be achieved in the subproblems.

However, it is appropriate to proceed a step further, by separating the problem both by scenarios and time periods. In this way, the number of subproblems increases from several hundreds to tens of thousands. From a parallel implementation point of view, there are several advantages to proceed in this way. It is possible to achieve a much better load balance among the processors; the smaller subproblems usually require solution times that are very similar, and any differences tend to be averaged out within a processor. Also, the solver for each subproblem does not have to take into account the sparsity of the problem data (note that the problem is nonlinear), and a straightforward implementation of a network solver suffices for the solution of the subproblems.

The decomposition by time periods requires a further step to be carried out. The variables representing the state of the system at the end of one period, that is, the reservoir levels at the end of each period must be separated between periods by including in the model additional variables, and constraints ensuring that the values of these variables are the same in the solution. These constraints are analogous to (42), i.e.

$$x^s = \overline{x}^s . \tag{45}$$

If they are also removed from the model through the use of Augmented Lagrangian terms, it results in a problem having constraints that can be separated by scenario and time period (and river system, if that is of interest).

In order to have a problem that is fully separable, it is necessary to deal with the quadratic penalty terms. Several approaches have been proposed in the literature to achieve this. For example, the quadratic terms can be replaced with proximal point terms, of the form

$$\frac{\rho}{2} \left\| x^s - \hat{x}^s \right\|^2 , \tag{46}$$

where $\hat{x}^s$ denotes the value of the variable from the preceding iteration. Alternatively, the quadratic term may be approximated by quadratic separable terms around the last computed point. A last approach is just to split the constraint into two,

$$x^s = z^s, \quad \overline{x}^s = z^s , \tag{47}$$

by adding extra parameters z^s, and then incorporating the constraints into Augmented Lagrangian terms and updating these parameters in some appropriate manner.

The algorithm implemented has the following structure:
- Step 0. Initialization.
    Initialize the values of variables, multipliers and parameters appropriately.
- Step 1. Subproblem solution.
    Solve each one of the subproblems corresponding to a scenario and time period, to a tolerance that is tightened with the iteration number. If the solution is carried out in a sequential manner,

update the parameters at the moment the information becomes available. For example, update the reference values z^s from the solution of the preceding subproblem.
- Step 2. Convergence checking.
  Check the termination conditions.
- Step 3. Updating.
  Update any parameters not yet modified, such as the multipliers, for example. Use a subgradient approach for this updating. Modify the termination tolerances for the subproblems.

The convergence results obtained are better if the updating mentioned in Step 1 is done as soon as possible. However, this may not be efficient in a distributed computation environment, due to transmission costs. In this case, part of the updating is carried out in Step 3.

The value of the penalty parameter can be adjusted throughout the algorithm.

The termination conditions are defined in terms of the satisfaction of the constraints, and more specifically of the value of the terms $\rho \| x^s - \hat{x}^s \|$. If these terms are sufficiently small, the constraints are satisfied, and the first-order conditions are also satisfied. The termination criteria cannot be too tight, due to the linear convergence of the procedure.

8.6.5 PARALLEL SOLUTION

The preceding algorithm is easy to implement in a distributed or parallel computation environment. The simplest approach is to use a master/slave scheme. In this scheme, one of the processors (the master) is in charge of (i) preparing the data to be distributed to the slave processors, (ii) controlling the solution of the subproblems, (iii) checking the optimality conditions and (iv) updating the parameters that depend on values computed by different slave processors.

The same processor where the master is running is also able to perform as a slave processor, due to the light tasks assigned to the master. The main concern in this approach is how to allocate the subproblems to the slave processors so that the load of these slave processors is balanced, and communication costs are reduced.

The procedures which can be implemented for the solution of the hydrothermal coordination problem can take advantage of the tree structure of the subproblems. By dividing the subproblems among the processors at a level as high on the tree as possible, the communication requirements between processors are minimized. On the other hand, a good load balance requires that the number of subproblems assigned to each processor is as similar as possible, provided that the processors are equivalent and they are lightly loaded. This may require introducing a larger number of subdivisions in the branches of the tree. It should be noted that the communication requirements between processors are limited. Only the information regarding the parameters and their updating, and the evaluation of the termination conditions, need to be exchanged between the processors. The main bulk of the information regarding variables and multipliers is only returned to the master at the end of the iteration process. These two conflicting goals should be properly balanced.

Additional details can be found in [13, 14].

8.7 References

1. Batut, J., Renaud, A. and Sandrin, P. (1990) New Software for the Generation Rescheduling in the Future EDF National Control Center. *Proceedings of the Tenth Power Systems Computation Conference, PSCC'90*, pp. 1163-1170. Graz, Austria.
2. Batut, J. and Renaud, A. (1992) Daily Generation Scheduling Optimization with Transmission Constraints: a New Class of Algorithms. *IEEE Transactions on Power Systems.* Vol. 7, No. 3, pp. 982-989.
3. Bazaraa M.S., Sherali H.D., and Shetty C.M. (1993) *Nonlinear Programming, Theory and Algorithms. Second Edition*, John Wiley and Sons, New York.
4. Benders, J.F. (1962) Partitioning Procedures for Solving Mixed-Variables Programming Problems. *Numerische Mathematik.* Vol. 4, pp. 238-252.
5. Bertsekas, D.P. (1982) *Constrained Optimization and Lagrange Multiplier Methods*, Academic Press, Cambridge.
6. Bertsekas, D.P., Lauer, G.S., Sandell, N.R. and Posbergh, T.A.. (1983) Optimal Short-Term Scheduling of Large-Scale Power Systems. *IEEE Transactions on Automatic Control.* Vol. 28, No. 1, pp. 1-11.
7. Brännlund, H., Sjelvgren, D. and Bubenko, J. (1988) Short Term Generation Scheduling with Security Constraints. *IEEE Transactions on Power Systems.* Vol. 1, No. 3, pp. 310-316.
8. Cohen, G. (1980) Auxiliary Problem Principle and Decomposition of Optimization Problems. *Journal of Optimization Theory and Applications.* Vol. 32, No. 3, pp. 277-305.
9. Conejo, A.J. and Aguado, J.A. (1998) Multi-Area Coordinated Decentralized DC Optimal Power Flow. *Paper PE-320-PWRS-0-12-1997.* Approved for publication in the IEEE Transactions of Power Systems.
10. Contaxis, G.C. and Kavatza, S.D. (1990) Hydrothermal Scheduling of a Multireservoir Power System with Stochastic Inflows. *IEEE Transactions on Power Systems*, Vol. PWRS-5, pp. 766-773.
11. Dantzig, G.B. and Infanger, G. (1992) Approaches to Stochastic Programming with Applications to Electric Power Systems. Lecture Notes on

Optimization in Planning and Operation of Electric Power Systems, SVOR/ASRO, Thun.
12. Dillon, T.S., Edwin, K.W., Kochs, H.D. and Tand, R.J. (1978) Integer Programming Approach to the Problem of Optimal Unit Commitment with Probabilistic Reserve Determination. *IEEE Transactions on Power Apparatus and Systems.* Vol. PAS-97, No. 6, pp. 2154-2166.
13. Escudero, L.F., de la Fuente, J.L., García, C. and Prieto, F.J. (1996) Hydropower Generation Management under Uncertainty via Scenario Analysis and Parallel Computation. *IEEE Transactions on Power Systems.* Vol. PWRS-11, pp. 683-689.
14. Escudero, L.F., de la Fuente, J.L., García, C. and Prieto, F.J. (1998) A Parallel Computation Approach for Solving Multistage Stochastic Network Problems. Approved for publication in the Annals of Operations Research.
15. Everett, H. (1963) Generalized Lagrange Multiplier Method for Solving Problems of Optimum Allocation of Resources. *Operations Research.* Vol. 11, pp. 399-417.
16. Ferreira, L.A.F.M. (1993) On the Duality Gap for Thermal Unit Commitment Problems. *ISCAS'93*, pp. 2204-2207.
17. Geoffrion, A.M. (1972) Generalized Benders Decomposition. *Journal of Optimization Theory and Applications.* Vol. 10, No. 4, pp. 237-260.
18. Hiriart-Urruty, J.B., and Lemaréchal C. (1996) *Convex Analysis and Minimization Algorithms. Vol. I and II.* Springer-Verlag, Belin.
19. Hobbs, W.J., Hermon, G., Warner, S. and Sheble, G.B. (1988) An Enhanced Dynamic Programming Approach for Unit Commitment. *IEEE Transactions on Power Systems.* Vol. 3, No. 3, pp. 1201-1205.
20. Jiménez Redondo, N. and Conejo, A.J. (1998) Short-Term Hydro-Thermal Coordination by Lagrangian Relaxation: Solution of the Dual Problem. *Paper PE-333-PWRS-0-12-1997.* Approved for publication in the IEEE Transactions of Power Systems.
21. Luh, P.B., Zhang, D. and Tomastik, R.N. (1997) An Algorithm for Solving the Dual Problem of Hydrothermal Scheduling. *IEEE Power Winter Meeting, Paper PE-601-PWRS-0-01-1997.* New York.
22. Luenberger D.G. (1989) *Linear and Nonlinear Programming. Second Edition,* John Wiley and Sons, New York.
23. Medina, J., Conejo A., Pérez Thoden F. and González del Santo J. (1994) Medium-Term Hydro-Thermal Coordination Via Hydro and Thermal Subsystem Decomposition. *TOP.* Vol. 2, No. 1, pp. 133-150.
24. Medina, J., Conejo, A.J., Jiménez Redondo, N., Pérez Thoden, F. and González, J. (1996) An LP Based Decomposition Approach to Solve the Medium-Term Hydro-Thermal Coordination Problem. *Proceedings of the Twelfth Power Systems Computation Conference, PSCC'96.* Vol. I, pp. 397-405. Dresden, Germany.

25. Medina, J., Quintana, V.H., Conejo, A.J. and Pérez Thoden, F. (1998) A Comparison of Interior-Point Codes for Medium-Term Hydro-Thermal Coordination. Accepted for publication in the IEEE Transactions on Power Systems.
26. Medina, J., Quintana, V.H. and Conejo, A.J. (1998) A Clipping-off Interior-Point Technique for Medium-Term Hydro-Thermal Coordination. *Paper PE-009-PWRS-0-1-1998*.Accepted for publication in the IEEE Transactions on Power Systems.
27. Mendes, V.M., Ferreira, L.A.F.M., Roldao, P. and Pestana, R. (1993) Optimal Short-Term Scheduling in Large Hydrothermal Power Systems. *Proceedings of the Eleventh Power Systems Computation Conference, PSCC'93*. Vol. II, pp. 1297-1303. Avignon, France.
28. Merlin, A. and Sandrin, P. (1983) A New Method for Unit Commitment at Electricité de France. *IEEE Transactions on Power Apparatus and Systems.* Vol. PAS-102, No. 5, pp. 1218-1225.
29. Muckstadt, J.A. and Koening, S.A. (1977) An Application of Lagrangian Relaxation to Scheduling in Power Generation Systems. *Operation Research.* Vol. 25, No. 3, pp. 387-403.
30. Pellegrino, F., Renaud, A. and Socroun, T. (1996) Bundle and Augmented Lagrangian Methods for Short-Term Unit Commitment. *Proceedings of the Twelfth Power Systems Computation Conference, PSCC'96.* Vol. II, pp.730-739. Dresden, Germany.
31. Pereira, M.V.F. and Pinto, L.M.V.G. (1991) Multi-Stage Stochastic Optimization Applied to Energy Planning. *Mathematical Programming.* Vol. 52, pp. 359-375.
32. Polyak B.T. (1987) *Introduction to Optimization*, Optimization Software, Inc, New York.
33. Rakic, M.V. and Marcovic, Z.M. (1994) Short Term Operation and Power Exchange Planning of Hydro-Thermal Power Systems. *IEEE Transactions on Power Systems.* Vol. 9, No. 1, pp. 359-365.
34. Renaud, A. (1993) Daily Generation Management at Electricité de France: from Planning towards Real Time. *IEEE Transactions on Automatic Control.* Vol. 38, No. 7, pp. 1080-1093.
35. Rockafellar, R.T. and Wets, R.J.-B. (1991) Scenario and Policy Aggregation in Optimization under Uncertainty. *Mathematics of Operations Research.* Vol. 16, pp. 119-147.
36. Rubinstein, R.Y. (1981) *Simulation and the Monte Carlo Method*, John Wiley and Sons, New York.
37. Sherkat, V.R., Campo, R., Moslehi, K. and Lo, E.O. (1985) Stochastic Long-Term Hydrothermal Optimization for a Multireservoir System. *IEEE Transactions on Power Apparatus and Systems*, Vol. PAS-104, pp. 2040-2050.
38. Wang, S.J. and Shahidehpour, S.M. (1993) Power Generation Scheduling for Multi-Area Hydro-Thermal Systems with Tie Line Constraints, Cascaded

Reservoirs and Uncertain Data". *IEEE Transactions on Power Systems*, Vol. PWRS-8, pp. 1333-1340.
39. Wang, S.J., Shahidehpour, S.M., Kirschen, D.S., Mokhtari, S. and Irisarri, G.S. (1995) Short-Term Generation Scheduling with Transmission and Environmental Constraints using an Augmented Lagrangian Relaxation. *IEEE Transactions on Power Systems*. Vol. 10, No. 3, pp. 1294-1301.
40. Wood, A.J. and Wollenberg, B.F. (1996) *Power Generation, Operation and Control. Second Edition*, John Wiley and Sons, New York.
41. Yan, H., Luh, P.B., Guan, X. and Rogan, P.M. (1993) Scheduling of Hydrothermal Power Systems. *IEEE Transactions on Power Systems*. Vol. 8, No. 3, pp. 1135-1365.
42. Zhuang, F and Galiana, F.D. (1988) Toward a More Rigorous and Practical Unit Commitment by Lagrangian Relaxation. *IEEE Transactions on Power Systems*. Vol. 3, No. 2, pp. 763-773.
43. Álvarez, M., Cuevas, C.M., Escudero, L.F., de la Fuente, J.L., García, C. and Prieto, F.J. (1994) Network Planning under Uncertainty with an Application to Hydropower Generation. *TOP*. Vol. 2, No. 1, pp. 25-58.

Chapter 9

INTERIOR POINT METHODS AND APPLICATIONS IN POWER SYSTEMS

Xie Kai And Y.H. Song
Department of Electrical Engineering and Electronics
Brunel University, Uxbridge
UB8 3PH, UK

9.1 Introduction

Interior-point methods (IPMs) are a central, striking feature of the constrained optimization landscape today. They have led a fundamental shift in thinking about continuous optimization. Today, in complete contrast to the era before 1984, researchers view linear and nonlinear programming from a unified perspective. The magnitude of this change can be appreciated simply by noting that no one would seriously argue today that linear programming is independent of nonlinear programming. Also, IPMs provide an alternative to active set methods for the treatment of inequality constraints, which permits the effective and efficient handling of large sets of equality and inequality constraints. Therefore, IPMs have been proposed for the solution of a wide range of traditional optimization problems in power systems since the 1990's, and the numerical experience with these methods has been quite positive.

This chapter first provides a brief tutorial on the recent development of IPMs. Karmarkar's projective method, logarithmic barrier function methods, the predictor-corrector variant of Primal-Dual method, reduced KKT system and IPMs for non-linear programming are described sequentially in section 2. Section 3 introduces IPM power system applications, and two of these, SCED (Security Constrained Economic Dispatching) and OPF (Optimal Power Flow), are taken to demonstrate the features of IPMs. Several critical implementation issues of a successful IPM are discussed in section 4 while the last section analyzes the numerical results of an IPM based spot-pricing algorithm in detail.

9.2 Interior Point Methods

9.2.1 BACKGROUND

The history of Interior Point Method can be traced back to 1955 [1]. Primarily in the form of barrier methods, interior point techniques were widely used during the 1960s to solve nonlinear constrained problems [2]. However during the 1970s, barrier methods were superseded by newly emerging, apparently more efficient alternatives such as augmented Lagrangian and sequential quadratic programming methods. By the early 1980s, barrier methods were almost universally regarded as a closed chapter in the history of optimization. On the other hand, although the first affine-scaling implementation of IPMs for linear programming was provided in 1964 [3], it could not challenge the total dominance of the Simplex method which was the only effective algorithm for linear programming for about 40 years, (starting from its discovery in 1947 until Karmarkar's breakthrough [4]), whose solution goes from corner point to corner point in the feasible set as indicated by x_i' in fig 1. The lack of success of IPMs before 1984 can be attributed to two reasons: First, due to the storage limitations, the size of the problems solved in the late 1960's never exceeded several hundred rows and columns and for such sizes, the Simplex method is practically unbeatable. Secondly, there was no sparse symmetric solver available at that time (they appeared at the beginning of 1970's) so the orthogonal projections must have killed the efficiency of IPMs. IPMs need significantly more memory than the Simplex methods.

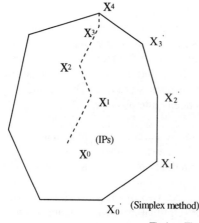

- Starting from an interior point, the method constructs a path that reaches the optimal solution after a few iterations
- Lead to a "good estimation" of the optimal solution after the few iterations. Very useful for Succesive LP.
- Always within feasible space
- The number of iterations are independent of the problem size

Fig 1   The Characteristics of IPMs

The picture changed dramatically in 1984. The current interest in IPMs was sparked by Karmarkar's paper [4], which used projective transformations to demonstrate a polynomial time complexity algorithm for linear programming. This paper and the other later papers claimed that the new method was significantly faster than existing implementations of the Simplex method. This captured the attention of the mathematical programming community, as indicated by the large number of papers published since 1984. Within two years of the publication of Karmarkar's method, an implementation of an IPM to solve the dual problem, which is became known as the dual affine variant [31] demonstrated overall superiority over a specific implementation of the Simplex method, MINOS 4.0. Shortly thereafter, similar performance was demonstrated for a primal-dual path following algorithm [32].

In 1986, Gill, Murry, Saunders, Tomlin and Wright [33] built a bridge between Karmarkar's method and Fiacco and McCormick's logarithmic barrier approach. They found that Karmarkar's method generates search directions which are similar to those generated by the *Newton method* applied to the first order conditions for a *Lagrangian* of an LP with a *logarithmic barrier function* - three elements of Primal-Dual IP methods (PDIPMs). Barrier methods were developed for the primal and for the dual LP formulation. This technique was hitherto applied to non-linear optimisation problems. In 1989, a Primal-Dual log-barrier algorithm set out by Mediddo [34], derived that the centre path is the locus of the successive optimal solution of the Lagrangian sub-problems. In 1992 Mehrotra developed predictor-corrector variant PDIPMs [35]. Nowadays, almost all the state-of-the-art IPMs implementations are those of Primal-Dual methods. Hence we concentrate only on Primal-Dual Methods in this chapter. Fig 1 shows some main characteristics of IPMs.

9.2.2 AFFINE SCALING METHODS

Karmarkar's original method had several properties: a special (non-standard) form was assumed for the linear programming; nonlinear projective geometry was used in its description; and no information was available about the implementation [4]. An early implementation difficulty with Karmarkar's method involved converting the standard form linear programming problem to Karmarkar's homogeneous form. And the inverse projective transformation needed to be converged in polynomial time. Essentially, Karmarkar's interior point method is based on two basic ideas: rescaling and projection into the nullspace. Therefore in practice, rescaling method is used to implement Karmarkar's projective method.

Geometrically, Karmarkar's method is a rescaling of the LP problem at each iteration step, thereby moving the iteration point into the centre of a symmetric polyhedron (bounded polytope). From this centre point it is easy to take a fairly long step along the projective gradient without running into difficulties with the boundaries of the feasible domain. For a standard linear programming problem

$$\text{Min } c^T x$$
$$\text{s.t. } Ax = b \quad (1)$$
$$x \geq 0$$

Where A is a m×n matrix, x and c are n vectors and b is m vector.

Start from a point inside the feasible set, suppose it is $x_0 = [1,1,\cdots 1]$ and move in a cost reducing direction. Since the cost is $c^T x$, the best direction is $-c$.

Normally, that takes us out of the feasible set, moving into that direction does not maintain $Ax = b$. If $Ax_0 = b$ and $Ax_1 = b$, then $\Delta x = x_1 - x_0$ has to satisfy $A\Delta x = 0$. The step Δx must be in the nullspace of A. Therefore we can define a projective transformation P to project $-c$ into the nullspace:

$$P = I - A^T(AA^T)^{-1}A \qquad (2)$$

P takes every vector into the nullspace because $AP = A - (AA^T)(AA^T)^{-1}A = 0$. If x is already in the nullspace, then $Px = x$ for $Ax = 0$. We don't compute the inverse $(AA^T)^{-1}$ in (2) directly to get P, instead we solve a linear equation:

$$(AA^T)y = Ac \qquad (3)$$

then $Pc = c - A^T y$. And $-Pc$ indicates the step Δx.

So far, we have completed the first idea—the projection which gives the steepest feasible decent. The second step needs a new idea, since repeating the same direction is useless.

Karmarkar's suggestion is to transform x_1 back to the centre position $e = [1,1,\cdots 1]$. That change of variables will change the problem, and then the second step projects c onto the nullspace of the new A. His changes of variables are non-linear, but the simpler transformation is just a rescaling of the axes. Rescale all vectors by $D^{-1}c$, D is a diagonal matrix defined as :

$$D = \begin{bmatrix} x_0^{(1)} & & \\ & x_0^{(2)} & \\ & & \ddots \end{bmatrix} \quad D^{-1}x = \begin{bmatrix} 1 \\ \vdots \\ 1 \end{bmatrix} = e \qquad (4)$$

We are back to at the centre with room to move. However, the feasible set has changed and so has the cost vector c. The rescaling from x to $X = D^{-1}x$ has two effects:

    the constraints $Ax = b$ becomes $ADX = b$,

    the cost $c^T x$ becomes $c^T DX$.

The new problem is to minimise $c^T DX$ subject to $ADX = b$ and $X \geq 0$ (which is equivalent to $x \geq 0$). The second step projects the new cost vector onto the nullspace of AD. Equation (3) changes to:

$$AD^2 A^T y = AD^2 c \text{ and then } PDc = Dc - DA^T y . (5)$$

Now you have the whole algorithm, except for starting and stopping. It could start from any feasible set. At each iteration step, the current guess x^k is rescaled to the point $e = [1,1,\cdots 1]$ and then the projection (5) gives the step direction in the rescaled variables.

The procedure of the Rescaling method can be summarised as:

Rescaling algorithm
(1) Construct a diagonal matrix D from the components of x^k. So that $D^{-1}x^k = e$
(2) With this D compute the projection PDc in equation (5)
(3) Determine the number s so that e-sPDc has a zero element
(4) Reduce that s by a factor α (say $\alpha = 0.96$)
(5) The new vector is $x^{k+1} = x^k - sDPDc$

All the work was in step 2 of this algorithm, which is a weighted projection. The weighted projection yields y, and its solution is a fundamental problem of numerical linear algebra. The normal way to compute y is elimination. That succeeds in a small problem, and also in a large one if all matrices are sparse. Notice (2), Karmarkar's original algorithm went beyond the rescaling $D^{-1}x$ which keeps the cost linear. Fig 2 shows an example of this rescaling transformation.

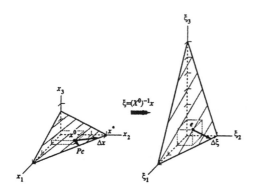

Fig 2 A step of rescaling algorithm

9.2.3 PRIMAL DUAL METHODS

Logarithmic barrier methods were introduced by Frisch [1] and developed by Fiacco and McCormick [2]. Initially, the concentration was on non-linear problems. The relationship between logarithmic barrier method and IPMs was first noted by Gill et al [33].

As in the projective method, the matrix $AD^2 A^T$ also plays an important role in a logarithmic barrier method. Every Newton step requires computing at least one orthogonal projection onto the null space of a scaled linear operator AD, where A is the LP constraint matrix, D is a positive diagonal scaling matrix that changes in subsequent iterations. Primal, Dual and Primal-Dual variants of IPMs differ in the way D is defined, but the effort to compute Karmarkar's projection is always the same. Every orthogonal projection involves inversion of the matrix $AD^2 A^T$ — the most time consuming linear algebra operation. After restructuring of the matrix, direct methods that compute a sparse symmetric factorisation can be used.

(1) Primal-Dual Log Barrier Methods

This algorithm can be easily derived by considering the first order conditions of the primal problem, or alternatively, by applying the logarithmic barrier method to the dual problem. Here the problem is:

$$\text{Maximise} \quad b^T y + \mu \sum_{j=1}^{n} \ln z_j \quad (6)$$

$$\text{Subject to} \quad A^T y + z = c$$

with the Lagrangian

$$L(x, y, z, \mu) = b^T y + \mu \sum_{j=1}^{m} \ln z_j - x^T (A^T y + z - c) \quad (7)$$

The first order conditions for (7) are

$$\begin{cases} Ax = b & \text{Primal feasibility} \\ A^T y + z = c & \text{Dual feasibility} \\ XZe = \mu e & \text{Compementarity condition} \end{cases} \quad (8)$$

Where $X = diag[x_i]$

$$Z = diag[c_i - a_i^T y]$$

The Newton method can be applied to (8) to yield the search direction

$$\Delta y = -(AXZ^{-1}A^T)^{-1}AZ^{-1}v(\mu)$$
$$\Delta z = -A^T\Delta y \qquad (9)$$
$$\Delta x = Z^{-1}v(\mu) - XZ^{-1}\Delta z$$

Where $v(\mu) = \mu e - XZe$.

In comparing primal, dual and primal-dual methods, we can notice that all need to construct a matrix of form ADA^T, where D is diagonal, and the content of D varies. But the computation work does not. Given this similarity, there are two immediate advantages that appear when examining the primal-dual method. The first is that for primal feasible x and dual feasible y and z, the exact current duality gap $c^Tx - b^Ty$ is always known, it can be easily shown that for a feasible point (x,y,z)

$$c^Tx - b^Ty = x^Tz \qquad (10)$$

thus an excellent measure of how close the given solution is to the optimal is always available. A second advantage is that it allows for separate step lengths in the primal and dual spaces, i.e.,

$$x^{k+1} = x^k + \alpha_P^k \Delta x^k$$
$$y^{k+1} = y^k + \alpha_D^k \Delta y^k \qquad (11)$$
$$z^{k+1} = z^k + \alpha_D^k \Delta z^k$$

This separate step algorithm has proven highly efficient in practice, significantly reducing the number of iterations to convergence [9].

The Primal-Dual search direction (9) is derived under the assumption of feasibility, if we do not assume that the point (x,y,z) is feasible, applying Newton's method to (9), yields the system

$$Z\Delta x + X\Delta z = \mu e - XZe$$
$$A\Delta x = b - Ax \qquad (12)$$
$$A^T\Delta y + \Delta z = c - A^Ty - z$$

Which has the solution

$$\Delta y = -(AXZ^{-1}A^T)^{-1}(AZ^{-1}v(\mu) - AXZ^{-1}\gamma_D - \gamma_P)$$
$$\Delta z = -A^T\Delta y + \gamma_D \qquad (13)$$
$$\Delta x = Z^{-1}v(\mu) - Z^{-1}X\Delta z$$

Where $\gamma_D = c - A^Ty - z \quad \gamma_P = b - Ax$

Clearly Newton's method can be applied in a similar fashion to the infeasible primal or dual method.

(2) Upper bounds:

In power system optimization, it is very common to have variables with upper or lower constraints (generator outputs, voltage magnitude etc.) Let us consider a primal linear programming problem with upper bounds to variables:

$$\text{Minimise } c^T x$$
$$\text{Subject to } Ax = b \qquad (14)$$
$$x + s = u \qquad x, s \geq 0$$

And its dual:

$$\text{Maximise } b^T y - u^T z$$
$$\text{Subject to } A^T y - t + z = c \qquad (15)$$
$$z, t \geq 0$$

Its Lagrangian:

$$L(x, s, y, \mu) = c^T x - y(Ax + b) - t(x + s - u) - \mu \sum_{i=1}^{n} \ln s_i - \mu \sum_{i=1}^{n} \ln x_i$$
$$(16)$$

The first order optimality conditions for (16) are:

$$Ax = b$$
$$x + s = u$$
$$A^T y + z - t = c$$
$$XZe = \mu e \qquad (17)$$
$$STe = \mu e$$
$$x > 0$$
$$s > 0$$

Where X, S, Z and T are diagonal matrices with the elements $x_i, s_i, z_i,$ and t_i respectively. μ is a barrier parameter and $z = \mu X^{-1} e$. The set of solutions of (17) ($x(\mu), s(\mu)$) and ($y(\mu), z(\mu), t(\mu)$) defines the central path of a primal and dual problem respectively. If some solution with a certain μ is available, then for the complementarity gap one has:

$$x^T z + s^T t = 2\mu e^T e = 2n\mu \qquad (18)$$

In any IPM, this quantity measures the error in complementarity. Observe that for feasible IPMs, the complementarity gap reduces to the usual duality gap and vanishes at an optimal solution.

(3) Reduced KKT system

A single iteration of the basic Primal-Dual algorithm makes one step of Newton's method applied to the first order optimality condition (17) with a given μ and then μ is updated (usually reduced). The algorithm terminates when infeasibility and complementarity gap is reduced below a predetermined tolerance. Newton's direction is obtained by solving the following system of linear equations:

$$\begin{bmatrix} A & 0 & 0 & 0 & 0 \\ I & 0 & I & 0 & 0 \\ 0 & A^T & 0 & I & -I \\ Z & 0 & 0 & X & 0 \\ 0 & 0 & T & 0 & S \end{bmatrix} \begin{bmatrix} \Delta x \\ \Delta y \\ \Delta s \\ \Delta z \\ \Delta t \end{bmatrix} = \begin{bmatrix} \xi_b \\ \xi_u \\ \xi_c \\ \mu e - XZe \\ \mu e - STe \end{bmatrix} \quad (19)$$

Where
$$\xi_b = b - Ax$$
$$\xi_u = u - x - s$$
$$\xi_c = c - A^T y - z + t$$

Let us now look closer at the system (19) after elimination

$$\Delta z = X^{-1}(\mu e - XZe - Z\Delta x)$$
$$\Delta s = \xi_u - \Delta x \quad (20)$$
$$\Delta t = S^{-1}(\mu e - STe - T\Delta s) = S^{-1}(\mu e - STe - T\xi_u + T\Delta x)$$

It reduces to
$$\begin{bmatrix} -D^{-2} & A^T \\ A & 0 \end{bmatrix} \begin{bmatrix} \Delta x \\ \Delta y \end{bmatrix} = \begin{bmatrix} \gamma \\ h \end{bmatrix} \quad (21)$$

$$D^2 = (X^{-1}Z - S^{-1}T)^{-1}$$

Where
$$\gamma = \xi_c - X^{-1}(\mu e - XZe) + S^{-1}(\mu e - STe) - S^{-1}T\xi_u$$
$$h = \xi_b$$

This matrix in the augmented system (21) is sparse symmetric but indefinite. The system of linear equations can be solved directly by Cholesky factorisation.

Another choice is to use the first set of (21) to solve for Δx, therefore the system can be reduced to a sparse, symmetric and positive definite normal equations systems:

$$(AD^2 A^T)\Delta y = AD^2 \gamma + h \qquad (22)$$

Both of the approaches have their own advantages. In the next section, we will point out the amazing connections between the reduced KKT system (21) and the augmented Hessian matrix in Optimal Power Flow (OPF). The matured super-sparsity techniques can be perfectly utilised. The advantages of this approach are more than just aesthetic. More generally, applying a sparsity preserving ordering scheme to the reduced KKT system offers great flexibility in the choice of orderings than if one works just with the matrix ADA^T. And in terms of numbers of arithmetic operations, the reduced system dominates the normal equation approach. A second advantage of the reduced system appears when one extends the interior point algorithm to quadratic programming problems. For quadratic programming, the same derivation can be repeated and the only change in (21) is that the Hessian Q of the quadratic objective function gets subtracted from $-D^{-2}$ in the upper left block of the symmetric matrix. If direction $\Delta x, \Delta s, \Delta y, \Delta z, \Delta t$ is computed, the maximum stepsize in it for primal α_p, and dual α_d, spaces that maintain non-negativity of variables are found and, often being slightly reduced by a factor α_0, a new iterate is computed:

$$\begin{aligned}
x^{k+1} &= x^k + \alpha_0 \alpha_P \Delta x \\
s^{k+1} &= s^k + \alpha_0 \alpha_P \Delta s \\
y^{k+1} &= y^k + \alpha_0 \alpha_D \Delta y \qquad (23) \\
z^{k+1} &= z^k + \alpha_0 \alpha_D \Delta z \\
t^{k+1} &= t^k + \alpha_0 \alpha_D \Delta z
\end{aligned}$$

(4) Algorithms of Primal-Dual Interior Point Method

Summarily, PDIP algorithm can be described as

Primal-Dual Interior Point algorithm
Input
$(x^0, s^0) > 0$ and $y^0, (z^0, t^0) > 0$ the initial pair of primal and dual solutions respectively
Parameters
ε is the accuracy parameter;
α_0 is the step size

Begin
$x = x_0; s = s_0; y = y_0; z = z_0; t = t_0$
While stop criteria is not satisfied, do calculate the search
Direction by
(21)(22)
calculate the new iterates by (23)
End
End

9.2.4 PREDICTOR-CORRECTOR METHOD

Computing $(\Delta x, \Delta y)$ from (21) or Δy from (22) is usually, when a direct method is applied, divided in two phases: factorisation of the matrix to some easily invertible form and the following solve that exploits this factorisation. Usually, the second step is at least an order of magnitude cheaper than the first step. This observation led to the introduction of higher order terms when computing direction. Computing different corrector terms resolves into multiple solution of the same linear system for several right-hand sides (it has been reduced by the same factorisation and is relatively inexpensive.)

Almost all interior point methods compose the direction step $\Delta x, \Delta s, \Delta y, \Delta z, \Delta t$ (denoted by Δ for short) from two parts

$$\Delta = \Delta_\alpha + \Delta_c \qquad (24)$$

i.e. combine the affine scaling, Δ_α, and centering, Δ_c, components. The term Δ_α is obtained by solving (19) for $\mu = 0$ and Δ_c is the solution of an equation like (19) with the right-hand side being

$$(0,0,0, \mu e - XZe, \mu e - STe)^T$$

Where $\mu > 0$ is the centering parameter ($\mu = (x^T z + s^T z)/(2n)$, for example, refers to centering that does not change the current complementarity gap). The term Δ_α is responsible for "optimisation", while Δ_c keeps the current iterate away from boundary.

Mehrotra proposed a predictor-corrector method, that can be derived directly form the first order conditions (17), by substituting $x + \Delta x, s + \Delta s, y + \Delta y, z + \Delta z, t + \Delta t$ into (17), it is then desired that a new estimate satisfies:

$$A(x+\Delta x) = b$$
$$(x+\Delta x) + (s+\Delta s) = u$$
$$A^T(y+\Delta y) + (z+\Delta z) - (t+\Delta t) = c \quad (25)$$
$$(X+\Delta X)(Z+\Delta Z)e = \mu e$$
$$(S+\Delta S)(T+\Delta T)e = \mu e$$

Collecting terms gives the system:
$$A\Delta x = b - Ax$$
$$\Delta x + \Delta s = u - x - s$$
$$A^T \Delta y + \Delta z - \Delta t = c - A^T y - z + t \quad (26)$$
$$Z\Delta x + X\Delta z = \mu e - XZe - \Delta X\Delta Ze$$
$$S\Delta t + T\Delta s = \mu e - STe - \Delta S\Delta Te$$

Where $\Delta x, \Delta z$ are n×n diagonal matrices with elements Δx_j and Δz_j, respectively. Examination shows that (26) is identical to (19) with the exception of the non-linear term, $\Delta X \Delta Z$ and $\Delta S \Delta T$ in (26). Corresponding to the two terms in (24), the solving of (26) can be divided into 2 steps:

(1) Predictor (Affine scaling direction)

Let $\mu = 0$ in (26), we have:

$$\begin{bmatrix} A & 0 & 0 & 0 & 0 \\ I & 0 & I & 0 & 0 \\ 0 & A^T & 0 & I & -I \\ Z & 0 & 0 & X & 0 \\ 0 & 0 & T & 0 & S \end{bmatrix} \begin{bmatrix} \Delta x_a \\ \Delta y_a \\ \Delta s_a \\ \Delta z_a \\ \Delta t_a \end{bmatrix} = \begin{bmatrix} \xi_b \\ \xi_u \\ \xi_c \\ -XZe \\ -STe \end{bmatrix} \quad (27)$$

By solving (27), we can find the maximum stepsize in the primal α_{Pa}, and in the dual α_{Da}, spaces preserving non-negativity of (x,s) and (z,t) and the predicted complementarity gap
$$g_a = (x + \alpha_{Pa} \cdot \Delta x_a)^T (z + \alpha_{Da} \Delta z_a) + (s + \alpha_{Pa} \cdot \Delta s_a)^T (t + \alpha_{Da} \Delta t_a) \quad (28)$$

is computed. It is then used to determine the barrier parameter μ:

$$\mu = \left(\frac{g_a}{g}\right)^2 \frac{g_a}{n} \tag{29}$$

Where $g = x^T z + s^T z$ denotes the complementarity gap. The term g_a / g measures the achievable progress in the affine scaling direction.

(2) Corrector

For such a μ in (29), the correction direction Δ_c is computed:

$$\begin{bmatrix} A & 0 & 0 & 0 & 0 \\ I & 0 & I & 0 & 0 \\ 0 & A^T & 0 & I & -I \\ Z & 0 & 0 & X & 0 \\ 0 & 0 & T & 0 & S \end{bmatrix} \begin{bmatrix} \Delta x_c \\ \Delta y_c \\ \Delta s_c \\ \Delta z_c \\ \Delta t_c \end{bmatrix} = \begin{bmatrix} 0 \\ 0 \\ 0 \\ \mu e - \Delta X_a \Delta Z_a e \\ \mu e - \Delta S_a \Delta T_a e \end{bmatrix} \tag{30}$$

and finally the direction Δ is determined. A single iteration of the (second order) Predictor-Corrector Primal-Dual method thus needs two solves of the same large, sparse linear system for two different right-hand side: first to solve (27) and then to solve (30).

We should notice here that the above Predictor-Corrector mechanism can be applied repeatedly, thus leading to a higher (than 2nd) order method. Although the use of higher order terms usually results in the reduction of the number of iterations, it does not necessary produce time saving due to the increased effort in a single iteration. To date, second-order methods seem, on average, to be the most efficient.

9.2.5 INTERIOR POINT METHODS FOR NON-LINEAR PROGRAMMING (NP) [40]

As we mentioned in section (9.2.1), it is apparent that IPMs for LP have a natural generalisation to the related field of convex non-linear optimisation which results in a new stream of research in applying IPMs for non-linear optimisation. For notational simplicity, we begin by considering the following non-linear programming problem:

$$\begin{aligned} & \text{Minimise } f(x) \\ & \text{Subject to } h_i(x) \geq 0 \qquad (i = 1, \cdots, m) \end{aligned} \tag{31}$$

Where x is a vector of dimension n and $f(x)$ and $h_i(x)$ are assumed to be twice continuously differentiable. This is a simplification of the general non-linear programming problem which can include equality constraints and bounds on variables. For the present, we consider this version of the problem, as it greatly simplifies the terminology, and the extension to the more general case is quite straightforward.

First add slack variables to each of constraints (31), reformulating the problem as:

$$\text{Minimise } f(x)$$
$$\text{Subject to } h_i(x) - w_i = 0 \quad (i = 1, \cdots, m)$$
$$w \geq 0 \quad (32)$$

Where h(x) and w represent the vectors with elements $h_i(x)$ and w_i respectively. We then eliminate the inequality constraints in (32) by placing them in a barrier term, resulting in the problem:

$$\text{Minimise } f(x) - \mu \sum_{i=1}^{m} \ln w_i$$
$$\text{Subject to } h_i(x) - w_i = 0 \quad (33)$$

Where the objective $L(x, w, \mu)$ is the classical Fiacco-McCormick logarithmic barrier function. The Lagrangian for this problem is:

$$L(x, w, \mu, y) = f(x) - \mu \sum_{i=1}^{m} \ln w_i - y^T (h(x) - w) \quad (34)$$

and the first order conditions for a minimum are:

$$\nabla_x L = \nabla f(x) - \nabla h^T(x) y = 0$$
$$\nabla_w L = -\mu W^{-1} e + y = 0 \quad (35)$$
$$\nabla_y L = h(x) - w = 0$$

Where W is the diagonal matrix with elements w_i. ∇h is the Jacobian matrix of the vector h(x). We now modify (35) by multiplying the second equation by W, producing the standard Primal-Dual system

$$\nabla f(x) - \nabla h^T(x) y = 0$$
$$-\mu e + WYy = 0 \quad (36)$$
$$h(x) - w = 0$$

Where again Y is the diagonal matrix with elements y_i.

The basis of the numerical algorithm for finding a solution to the Primal-Dual system (36) is Newton's method, which is well known to be very efficient. To simplify notation and at the same time highlight the connection with linear and quadratic programming, we introduce the following definitions:

$$H(x,y) = \nabla^2 f(x) - \sum_{i=1}^{m} y_i \nabla^2 h_i(x) \text{ and } A(x) = \nabla h(x)$$

The Newton's system for (36) is then:

$$\begin{bmatrix} H(x,y) & 0 & -A(x)^T \\ 0 & Y & W \\ -A(x) & I & 0 \end{bmatrix} \begin{bmatrix} \Delta x \\ \Delta w \\ \Delta y \end{bmatrix} = \begin{bmatrix} -\nabla f(x) + A(x)^T y \\ \mu e - WYe \\ h(x) - w \end{bmatrix} \quad (37)$$

The system (37) is asymmetric, but is easily symmetrised by multiplying the second equation by W^{-1}, yielding:

$$\begin{bmatrix} H(x,y) & 0 & -A(x)^T \\ 0 & W^{-1}Y & I \\ -A(x) & I & 0 \end{bmatrix} \begin{bmatrix} \Delta x \\ \Delta w \\ \Delta y \end{bmatrix} = -\begin{bmatrix} \delta \\ \gamma \\ \rho \end{bmatrix} \quad (38)$$

Where

$$\delta = \nabla f(x) - A(x)^T y \quad (39)$$
$$\gamma = y - \mu W^{-1} e \quad (40)$$
$$\rho = w - h(x) \quad (41)$$

Note that ρ measures primal infeasibility, by analogy with linear programming, we refer to δ as dual infeasibility.

We solve (38) or a minor modification of it, to find the search direction $\Delta x, \Delta y, \Delta w$, since the second equality can be used to eliminate Δw without producing any off-diagonal fill-in in the remaining system, recalling the reduced system in section (9.2.3), we can normally perform this elimination first, hence Δw is given by:

$$\Delta w = WY^{-1}(-\gamma - \Delta y)$$

and the resulting reduced KKT system is given by:

$$\begin{bmatrix} H(x,y) & -A(x)^T \\ -A(x) & -WY^{-1} \end{bmatrix} \begin{bmatrix} \Delta x \\ \Delta y \end{bmatrix} = \begin{bmatrix} -\delta \\ -\rho + WY^{-1}\gamma \end{bmatrix} \quad (42)$$

The algorithm then proceeds iteratively from an initial point x^0, y^0, w^0 through a sequence of points determined iteratively using the above calculated above step direction:

$$x^{k+1} = x^k + \alpha_P^k \Delta x^k$$
$$y^{k+1} = y^k + \alpha_D^k \Delta y^k \qquad (43)$$
$$w^{k+1} = w^k + \alpha_P \Delta w^k$$

9.3 Power System Applications

To the best of our knowledge, IPMs were first introduced into power systems research in the late 1980' for state estimation. In the early 1990's, IPMs were beginning to attract more and more power system scholars. A large number of papers, in a considerable variety of fields, such as security constrained economic dispatching, optimal reactive dispatching, optimal power flow, optimal reservoir dispatching, voltage stability and preventive control, which resulted in a new stream of research. Table one summarises the power system applications of IPMs.

Table 1 Interior Point Methods Applications in Power Systems

Application	Methods	Affine Scaling	Primary Dual for LP or QP	Primary Dual for NLP
State Estimation		Clements (1991) [10]		Wei (1998) [43]
Generator Scheduling	Reactive Power Optimization			Granville(1994)[14] Martinez (1996)[17]
	Economic Dispatching	Vargas (1993)[12] Momoh(1994)[15]	Yan (1997) [13] Irisarri (1998) [11]	
	Optimal Power Flow			Wu (1994) [18] Wei (1998) [43] Quitana (1995) [37]
Voltage Stability	Power Flow Unsolvability			Granville (1996) [19]
	Maximum Loadability			Irisarri(1997) [21] Wang (1998) [44]
	Simultaneous Transfer Capacity			Mello (1996) [20]
Hydro-Thermal Co-ordination				Wei (1998) [46]
Network Constrained Security Control		Lu (1993) [23]		
Spot Pricing				Xie (1998) [22]

In the application of interior point methods to power system optimisation, two basic strategies are generally reported in the literature. The first is based on a load flow optimisation scheme where the IPMs are applied to the resulting linear or quadratic programming problems obtained form the linearisation of power flow equations during the solution of the load flow algorithm [12][13][23]. Another is to apply the IPMs directly to the original non-linear programming problem [14][17]-[22]. The second strategy does not depend on the convergence of any power flow algorithm iterative scheme, and the power flow equations are only required to be attained at the optimal solution. Also numerical experiences have shown that direct methods are very effective in dealing with the large scale, ill-conditioned and voltage problem networks. In this section, SCED (Security Constrained Economic Dispatching) and OPF (Optimal Power Flow) problems are highlighted to demonstrate these two schemes.

9.3.1 SECURITY CONSTRAINED ECONOMIC DISPATCHING (SCED)

Given the generation committed to operate during each hour of a utility's operation or planning horizon, the SCED problem deals with the optimal economic allocation of this generation subject to transmission and generation constraints. The solution of this problem considers two major aspects: network modelling and mathematical methods to solve the optimisation problem. Many different optimisation methods have been proposed to solve the SCED problem, however the introduction of on-line functions in modern energy control centres, together with the growth of network size, demand faster and more reliable numerical techniques to solve the optimisation problems. References [12][13] linearised the original optimal problem and a DC power flow was used to obtain the Generalised Generation Distribution Factor (GGDF) and Incremental Transmission Loss Factor (ITLF), and IPM was employed to solve the yielded successive linear programming problem. In [11], the SCED model accounted for ramping rate constraints and was solved by a quadratic programming interior point method.

In a SCED, the real-power generations are controlled to minimise the total generation cost subject to power balance, transmission and generation constraints, while the reactive power controls are assumed constant. Mathematically, the SCED is a non-linear programming problem that can be formulated as:

$$\text{Min}\left\{ C_i = \sum_{i \in G} C_i(P_i) \right\} \quad (44a)$$

$$\text{s. t.} \sum_{i \in G} P_i = P_D + P_L \quad (44b)$$

$$P_{i\min} \le P_i \le P_{i\max} \quad (44c)$$

$$F_{l\min} \leq |F_l| \leq F_{l\max} \tag{44d}$$

Where P_i is the active power injection of generator i;

F_l is the active power flow through line l;

$C_i(P_i)$ is the generation cost curve of generator i, second order representation is adopted here.

(1) Linear programming format

Linearise the above problem under the following assumption

- To relate the active flows with the generation power (GGDF)

$$F_l = \sum_{i \in G} \beta_{i,l} P_i \tag{45}$$

$\beta_{i,l}$ is the GGDF that relates the active power flow through line l with the generation i.

- A small variation in the generating units produces a change in the losses of the system if the load is constant

$$\Delta P_L = \sum_{i \in G} P_i \tag{46}$$

- To relate the active power losses with the generation power Incremental Transmission Losses Factors (ITLF)

$$\Delta P_L = \sum_{i \in G} \gamma_i \Delta P_i \tag{47}$$

γ_i is the ITLF that relates the losses with the real power generation bus i, and it is updated from the current AC power flow solution at each iteration.

By using these assumptions, the linearised SCED problem at the rth iteration becomes:

$$\text{Min} \left\{ \Delta C_t^r = \sum_{i \in G} (b_i + 2c_i P_i^r) \Delta P \right\} \tag{48}$$

$$\text{s.t.} \sum_{i \in G} (1 - \gamma_i^r) \Delta P_i = 0 \tag{49}$$

$$\Delta P_{i\min} \leq \Delta P_i \leq \Delta P_{i\max} \tag{50}$$

$$\Delta F_{l\min} \leq \sum_{i \in G} \beta_{i,l} \Delta P_i \leq \Delta F_{l\max} \tag{51}$$

Where $\Delta P_{i\min}$ and $\Delta P_{i\max}$ are the step bounds on ΔP_i, and can be defined as: $\Delta P_{i\max} = P_{i\max} - P_i^r$ and $\Delta P_{i\min} = P_{i\min} - P_i^r$ respectively. $\Delta F_{i\min}$ and $\Delta F_{i\max}$ are the step bounds on ΔF_i, $\Delta F_{i\max} = F_{i\max} - P_i^r$ and $\Delta F_{i\min} = F_{i\min} - P_i^r$.

Obviously, equations (48~51) comply with the standard LP formulation except for the "security constraints": $\sum_{i \in G} \beta_{i,l} \Delta P_i \leq \Delta F_{l\max}$. Such a constraint can be converted to an equality constraint by introducing a new variable ΔP_l:

$$\sum_{i \in G} \beta_{i,l} \Delta P_i + \Delta P_l = \Delta F_{l\max}$$
$$\Delta P_{l\min} \leq \Delta P_l \leq \Delta P_{l\max} \quad (52)$$

Therefore, (48~52) construct the exact form of standard LP (1) which can be solved for ΔP_i, and then the value of P_i is updated: $P_i^{r+1} = P_i^r + \Delta P_i$. And the stop criterion for the SLP problem is :

$$\frac{|C(P_G^r) - C(P_G^{r-1})|}{|C(P_G^r)|} \leq \delta \quad (53)$$

For every single LP sub-problem, the IPM is employed in obtaining a solution. The crucial part of Fig.2.1 is to solve LP problems via IPMs. In [12], a dual-affine method mentioned in section 1.3.1 was employed, [11][15] utilized quadratic programming technique and [13] compared Primal-Dual Interior Point method and Predictor-Corrector Primal-Dual Interior Point method (section 9.2.3 and 9.3.4). The results, from the 236-bus to 2124-bus systems that the authors tested, showed the PCPDIP converged much faster than then the pure PDIP, taking only 40%-50% of the numbers of iterations of the latter.

(2) Quadratic programming format and ramping constraints

Many procedures for economic dispatching attempt to deal with the ramping rate constraints, an time-separated constraint, which is important for rationally allocating generation capacity along the time horizon. With this in mind, the SCED problem has to be formulated as a time-dependent problem. Assuming the generator cost curves are quadratic, then the cost function may be stated as:

$$\text{Min} \sum_{t=1}^{nt} (\frac{1}{2}p(t)^T Q(t) p(t) + c(t)^T p(t))$$

$$\text{s.t.} \sum_{i \in G} P_i(t) = P_D(t) + P_L(t) = d(t)$$

$$P_{i\min}(t) \le P_i(t) \le P_{i\max}(t) \qquad (54)$$

$$F_{l\min}(t) \le |F_l(t)| \le F_{l\max}(t)$$

$$R_{i\min}(t) \le P_i(t+1) - P_i(t) \le R_{i\max}(t)$$

Where nt is the number of time intervals. The last equation in (54) is ramping constraint. Define

$$G(t) = \begin{pmatrix} -I \\ I \\ \beta \\ -\beta \end{pmatrix}, \; g(t) = \begin{pmatrix} -P_{\min}(t) \\ P_{\max}(t) \\ F_{\max} \\ -F_{\min} \end{pmatrix}$$

$$\beta^T = [\beta_1^T \cdots \beta_{nl}^T]$$

$$F_{\max} = [F_{1\max}, \cdots F_{nl\max}], \; m(t) = \begin{bmatrix} R(t) \\ -R(t) \end{bmatrix}$$

$$F_{\min} = [F_{1\min}, \cdots F_{nl\min}]$$

The dispatch problem can now be written

$$\text{Min} \quad \frac{1}{2} P^T Q P + c^T P$$

Subject to:

$$EP = d$$
$$GP \le g \qquad (55)$$
$$MP \le m$$

Where P,d,g,m are vectors of time-dependant; e.g., $P^T = ((P(1)^T \cdots P(nt)^T)$; Q, E, G are block diagonal matrices of time-dependant sub-matrices; e.g. $Q = diag(Q(1), \cdots Q(nt))$ and

$$M = \begin{bmatrix} -I & I & 0 & \cdots & 0 \\ I & -I & 0 & \cdots & 0 \\ 0 & -I & I & \cdots & 0 \\ 0 & I & -I & \cdots & 0 \\ \vdots & & & \ddots & \vdots \\ 0 & 0 & 0 & \cdots & -I \end{bmatrix}$$

Equations (55) construct a quadratic programming problem.

(3) Interior Point implementation

Following the Primal Dual Interior Point method (9.2.3), the inequality constraints appearing in (55) are converted to equality constraints by the addition of a vector of slack variables:

$$GP + s_1 = g \\ MP + s_2 = m \tag{56}$$

With the introduction of a logarithmic barrier function, the Lagrangian for the SCED problem is thus

$$L = \frac{1}{2} P^T QP + c^T P - \lambda^T (EP - d) + \pi_1 (GP + s_1 - g) \\ + \pi_2 (MP + s_2 - m) - \mu (\sum_{i=1}^{n1} \ln s_1 + \sum_{i=1}^{n2} \ln s_2) \tag{57}$$

Where λ, π_1 and π_2 are Lagrange multipliers.

The KKT necessary conditions require that the partial derivatives of the Lagrangian vanish at optimality

$$\nabla_P L = QP + c - E^T \lambda + G^T \pi_1 + M^T \pi_2 = 0 \\ \nabla_\lambda L = -EP + d = 0 \\ \nabla_{\pi_1} L = GP + s_1 - g = 0 \\ \nabla_{\pi_2} L = MP + s_2 - m = 0 \\ \nabla_{s_1} L = \pi_1 - \mu [S_1]^{-1} e = 0 \\ \nabla_{s_2} L = \pi_2 - \mu [S_2]^{-1} e = 0 \tag{58}$$

A Newton iteration is constructed by replacing $P, \lambda \cdots \cdots s_2$ with $P + \Delta P$, $\lambda + \Delta \lambda$, $\cdots \cdots s_2 + \Delta s_2$, and discarding all second order terms. Recall the

KKT system derived in section (9.2.3), eliminate all slack variables, the resulting system has the form

$$Q\Delta P - E^T \Delta \lambda + G^T \Delta \pi_1 + M^T \Delta \pi_2 = -\nabla_p L$$
$$-E\Delta P = -\nabla_\lambda L$$
$$G\Delta P - [\pi_1]^{-1}[S_1]\Delta \pi_1 = -\nabla_{\pi 1} L - \mu[\pi_1]^{-1} e + s_1 \quad (59)$$
$$M\Delta P - [\pi_2]^{-1}[S_2]\Delta \pi_2 = -\nabla_{\pi 2} L - \mu[\pi_2]^{-1} e + s_2$$

Where $[\pi_i], [S_i]$ (i=1,2) are diagonal matrices with the diagonal entries π_i, and s_i respectively.

All the matrices in (59) are block diagonal with the exception of M. As a result, if the ramp rate were not present, the equations would be decoupled in time and could be solved independently for each time period. As a consequence, if the variables and equations are reordered, grouping variables and equations corresponding to a given time period together, the resulting system has a bordered block diagonal structure.

$$\begin{bmatrix} K_1 & \cdots & & N_1^T \\ \vdots & \ddots & & \vdots \\ 0 & \cdots & K_{nt} & N_{nt}^T \\ N_1 & \cdots & N_{nt} & D \end{bmatrix} \begin{bmatrix} \Delta z_1 \\ \vdots \\ \Delta z_{nt} \\ \Delta \pi_2 \end{bmatrix} = \begin{bmatrix} b_1 \\ \vdots \\ b_{nt} \\ b_{\pi_2} \end{bmatrix} \quad (60)$$

Where $\Delta z_t = (\Delta P(t), \Delta \lambda(t), \Delta \pi_1(t))$, N_t is the rearrangement of $M(t)$ padded with columns of zeros such that $\sum_t N_t \Delta z_t = M\Delta P$ in the last set of equations of (59), and K_t is the coefficient matrix of Δz_t in the remaining equations of (59). This bordered block diagonal structure is exploited in implementing the iterative Newton solution. b_t is the rearrangement of right hand side terms of (59).

The crucial step in the solution of (60) is to factor each of the diagonal blocks as $K_t = L_t D_t L_t^T$, where L_t is lower triangular and D_t is diagonal. Then write $N_t K_t^{-1} N_t^T = J_t^T D_t^{-1} J_t$, where J_t is formed column by column using forward substitution in $L_t J_t = N_t$. $K_t^{-1} b_t$ can be obtained by forward-backward substitution also. Eliminating down the diagonal of (70), $\Delta \pi_2$ can be solved via the following equations

$$\left[D - \sum_{t=1}^{nt} N_t K_t^{-1} N_t^T\right] \Delta \pi_2 = b_{\pi_2} - \sum_{t=1}^{nt} N_t K_t^{-1} b_t \qquad (61)$$

Substitute $\Delta \pi_2$ into (60), $\Delta z_t, t = 1 \cdots nt$, can be solved.

9.3.2 OPTIMAL POWER FLOW (OPF)

(1) Introduction

The principal goal of an optimal power flow program is to provide the electric utilities with suggestions (setpoints) to optimize the current power system state online with respect to various objectives such as minimization of production costs or system active power losses, maximize the degree of security of a network or even a combination of some of these. The achievement of these goals are important to utilities, since often they are obliged by law to operate the network with consumption of minimal resources and a maximum degree of security. OPF consists of three parts: The set of equality constraints representing the power system model for static computations, the set of inequality constraints representing real-world and practical operational constraints whose violation is not acceptable in the power system, and the objective function. The major difference between OPF and SCED is the network modeling. OPF incorporates a complete model of power network, including both active and reactive power. Since it was defined by Carpentier [39], it has been attracting many researchers as a potentially powerful tool for power system operators and planners. Over the last three decades various optimization methods have been applied to this fascinating topic [41]. Essentially, OPF is a non-linear optimization problem. The landmark paper by Sun et al. [24] introduced into OPF problem a combination of the Newton's method with a Lagrangian multiplier method and penalty functions. Its well designed data structure and efficient use of sparsity techniques has become a standard in power system application. However, the major difficulty turned out to be the efficient identification of binding inequalities, e.g. to find out the active set, and the handling of functional inequality constraints (such as line flow). As we have seen in 9.2.5, it is natural to extend IPM to NP problems. When IPMs are extended to OPF[14][18], it is found that the two problems haunting Newton OPF can be solved perfectly. Ref. [42] utilizes a reduced correction equation to handle inequality constraints, which makes the size of the problem depend on the equality constraints via perturbed KKT condition and in rectangular form. Ref. [22] achieves the same result by applying polar form and Primal Dual IPM.

(2) Problem Formulation

For the production cost minimization problem, the OPF problem can be summarized as the following nonlinear programming problem

$$\min \; F(z)$$
$$s.t. \; g(z) = 0 \qquad (62)$$
$$h_l \leq h(z) \leq h_u$$

Objectives:

Both active and reactive power production costs are accounted for in the objective function and simulated by quadratic curves.

$$Min \; F = \sum_{i=1}^{m} (f_{pi}(P_i) + f_{qi}(Q_i)) \qquad (63)$$

Where $f_{pi}(P_{gi}) = CP_i^2 + BP_i + A$, $f_{qi}(Q_{gi}) = C_r Q_i^2 + B_r Q_i + A_r$

Equality constraints $g(z)$ include:

Real power balance

$$g_{P_i} = P_i - P_{di} - \sum_{j=1}^{n} V_i V_j |Y_{ij}| \cos(\theta_i - \theta_j - \delta_{ij}) = 0 \qquad (64)$$

Reactive power balance

$$g_{q_i} = Q_i - Q_{di} - \sum_{j=1}^{n} V_i V_j |Y_{ij}| \sin(\theta_i - \theta_j - \delta_{ij}) = 0 \qquad (65)$$

Inequality constraints h(z) include:

Generator capacity limits

$$P_{i\min} \leq P_i \leq P_{i\max} \qquad (66)$$
$$Q_{i\min} \leq Q_i \leq Q_{i\max} \qquad (67)$$

Voltage limits

$$V_{i\min} \leq V_i \leq V_{i\max} \qquad (68)$$

Transmission congestion limits

$$F_{l\min} \leq F_l \leq F_{l\max} \qquad (69)$$

Spinning reserve

$$R_i = \begin{cases} P_{i\max} - P_i & \text{if } P_i \geq (1 - f_i) P_{i\max} \quad \text{(Type 1)} \\ f_i P_{i\max} & \text{if } P_i < (1 - f_i) P_{i\max} \quad \text{(Type 2)} \end{cases} \qquad (70)$$

$$\sum_i R_i \geq R_{\min} \qquad (71)$$

Where f_i is a factor representing the maximum percentage of the capacity of generator i available for spinning reserve, as pledged by generator i or by system regulation. The curtailable load can also be viewed as another type of spinning reserve.

(3) Interior Point Implementation

Translate inequality constraints into equality constraints through introducing slack variables, and construct Lagrangian function:

$$\min L = F(z) - \lambda^T g(z) + \pi_l^T (h(z) - s_l - h_l)$$
$$+ \pi_u^T (h(z) + s_u - h_u) - \mu(\sum_{i=1}^{k} \ln s_{li} + \sum_{i=1}^{k} \ln s_{ui}) \qquad (72)$$

where $s_l \geq 0, s_u \geq 0$ are slack variables, $\pi_l < 0, \pi_u > 0$ are Lagrangian multipliers and, $\mu > 0$ is the barrier factor.

The KKT first order optimality conditions give:

$$\nabla_z L = \nabla_z f(z) - J^T(z)\lambda + \nabla_z h^T (\pi_l + \pi_u) = 0 \qquad (73)$$
$$\nabla_{s_l} L = [s_l]\pi_l + \mu * e = 0 \qquad (74)$$
$$\nabla_{s_u} L = [s_u]\pi_u - \mu * e = 0 \qquad (75)$$
$$\nabla_\lambda L = g(z) = 0 \qquad (76)$$
$$\nabla_{\pi_u} L = h(z) - s_l - h_l = 0 \qquad (77)$$
$$\nabla_{\pi_u} L = h(z) + s_u - h_u = 0 \qquad (78)$$

Where $[s_l] = diag(s_{l1}, s_{l2}, \cdots, s_{lk}), [s_u] = diag(s_{u1}, s_{u2}, \cdots, s_{uk})$

Using the Newton method to solve the above non-linear equations (73)~(78), we have:

$$H(z,\lambda)\Delta z - J^T(z)\Delta\lambda + \nabla_z h^T \Delta\pi_l + \nabla_z h^T \Delta\pi_u$$
$$= -(\nabla_z f(z) - J^T(z)\lambda + \nabla_z h^T (\pi_l + \pi_u)) \qquad (79)$$
$$- J(z)\Delta z = -g(z) \qquad (80)$$
$$[s_l]\Delta\pi_l + [\pi_l]\Delta s_l = -[s_l]\pi_l - \mu e \qquad (81)$$
$$[s_u]\Delta\pi_u + [\pi_u]\Delta s_u = -[s_u]\pi_u + \mu e \qquad (82)$$
$$\nabla_z h \Delta z - \Delta s_l = -(h(z) - s_l - h_l) = 0 \qquad (83)$$
$$\nabla_z h \Delta z - \Delta s_u = -(h(z) + s_u - h_u) \qquad (84)$$

Where $H(z,\lambda) = \nabla_z^2 f(z) - \sum_{i=1}^{2m} \lambda_i \nabla_z^2 g(z) + \sum_{i=1}^{k} (\pi_{ui} + \pi_{li})\nabla_z^2 h_i(z)$

$$(85)$$

Finally solving the above equations (79~84), we have:

$$\Delta s_l = \nabla_z h \Delta z \qquad (86)$$

$$\Delta s_u = -\nabla_z h \Delta z \qquad (87)$$

$$\Delta \pi_l = -[s_l]^{-1}([s_l]\pi_l + \mu e) - [s_l]^{-1}[\pi_l]\nabla_z h \Delta z \qquad (88)$$

$$\Delta \pi_u = -[s_u]^{-1}([s_u]\pi_u - \mu e) + [s_u]^{-1}[\pi_u]\nabla_z h \Delta z \qquad (89)$$

$$\begin{bmatrix} \Delta z \\ \Delta \lambda \end{bmatrix} = -A^{-1} \begin{bmatrix} \tilde{t} \\ g(z) \end{bmatrix} \qquad (90)$$

Where:

$$A = \begin{bmatrix} \tilde{H}(z,\lambda) & -J^T(z) \\ -J(z) & 0 \end{bmatrix} \qquad (91)$$

$$\tilde{H}(z,\lambda) = H(z,\lambda) + \nabla_z h^T(-[s_l]^{-1}[\pi_l] + [s_u]^{-1}[\pi_u])\nabla_z h \qquad (92)$$

$$\tilde{t} = \nabla_z f(z) - J^T(z)\lambda$$
$$+ \nabla_z h^T(\pi_l + \pi_u - [s_l]^{-1}([s_l]\pi_l + \mu e) - [s_u]^{-1}([s_u]\pi_u - \mu e))$$
$$\qquad (93)$$

Equation (90) is a large linear system which is also a reduced KKT system. Most of the computational effort is to format and factor the symmetric matrix A.

(4) Sparsity Techniques

The main problem in the OPF formulation is the overwhelming size of the problem. Notice A is the extended formulation of augmented Hessian matrix in Newton OPF[24]. All the influences of inequality constraints have been introduced into A directly, which avoids the active set determination procedure of Newton OPF. Because slack variables and dual variables do not appear in (90), A has the same dimension with its counterpart in Newton OPF, which is only determined by the equality constraints. Therefore, the blocking and factoring strategy of Newton OPF can be utilized in the implementation of PDIP conveniently and efficiently. If h(z) is the variable type constraint (P, Q, V etc.), It only needs to change the corresponding diagonal elements of $\tilde{H}$ and right hand side vector. However, if h(z) includes functional type constraints (line flow, real power reserve and other security constraints, etc.), it will be more complicated, non-diagonal elements of $\tilde{H}$ should be changed as well. Fortunately, because all the elements of $\nabla_z h$ and $\nabla_z^2 h$ are the byproducts of Newton OPF, the calculation burden of this modification is very limited. The augmented Hessian matrix A in (91) can be rearranged into the following special block substructure to make the matrix relatively easy to factor:

$$A = \begin{bmatrix} G & I^T \\ I & H \end{bmatrix} \quad (94)$$

Where G is a diagonal matrix, the handling of G is usually independent of the handling of H in Newton OPF. I is a matrix consisting 0's and 1's. The elements of G and H should be modified, only constraints of spinning reserve can effect the structure of G. If all units belong to type 1, except for unit 2, the modified G is as follows:

$$G = \begin{bmatrix} \frac{\partial^2 L}{\partial P_{g1}^2} + \delta & \delta & & \delta & & & \\ & \frac{\partial^2 L}{\partial P_{g2}^2} & & & & & \\ \delta & & \frac{\partial^2 L}{\partial P_{g3}^2} + \delta & \delta & & & \\ \delta & & \delta & \frac{\partial^2 L}{\partial P_{g4}^2} + \delta & & & \\ & & & & \ddots & & \\ & & & & & \frac{\partial^2 L}{\partial Q_{gi}^2} & \\ & & & & & & \ddots \end{bmatrix} \quad (95)$$

H should be modified to satisfy line flow constraints. If kl if the number of nodes connected to node i, the corresponding H_{ii}, H_{ij} would be in the forms of equations (96) and (97) where the items with superscript 0 represent the associated elements in the pre-updated Because H_{ii} is symmetric, only upper triangle entries are presented.

$$H_{ii} = \begin{bmatrix} \frac{\partial^2 L}{\partial \vartheta_i^2}^0 + \sum_{j=1}^{kl} \delta_{lj} \left(\frac{\partial P_{ij}}{\partial \vartheta_i} \right)^2 & \frac{\partial^2 L}{\partial \vartheta_i \partial V_i}^0 + \sum_{j=1}^{kl} \delta_{lj} \left(\frac{\partial P_{ij}}{\partial \vartheta_i} \frac{\partial P_{ij}}{\partial V_i} \right) & \frac{\partial^2 L}{\partial \vartheta_i \partial \lambda_{p3}}^0 & \frac{\partial^2 L}{\partial \vartheta_i \partial \lambda_{q3}}^0 \\ & \frac{\partial^2 L}{\partial V_i^2} + \sum_{j=1}^{kl} \delta_{lj} \left(\frac{\partial P_{ij}}{\partial V_i} \right)^2 & \frac{\partial^2 L}{\partial V_i \partial \lambda_{pi}} & \frac{\partial^2 L}{\partial V_i \partial \lambda_{q3}} \\ & & 0 & 0 \\ & & & 0 \end{bmatrix}$$
(96)

$$H_{ij} = \begin{bmatrix} \dfrac{\partial^2 L}{\partial \vartheta_i \partial \vartheta_j}^0 + \delta_{lj} \dfrac{\partial P_{ij}}{\partial \vartheta_i} \dfrac{\partial P_{ij}}{\partial \vartheta_j} & \dfrac{\partial^2 L}{\partial \vartheta_i \partial V_{ji}}^0 + \delta_{lj} \dfrac{\partial P_{ij}}{\partial \vartheta_i} \dfrac{\partial P_{ij}}{\partial V_j} & \dfrac{\partial^2 L}{\partial \vartheta_i \partial \lambda_{pj}}^0 & \dfrac{\partial^2 L}{\partial \vartheta_i \partial \lambda_{qj}}^0 \\[6pt] \dfrac{\partial^2 L}{\partial V_i \partial \vartheta_j}^0 + \delta_{lj} \dfrac{\partial P_{ij}}{\partial V_i} \dfrac{\partial P_{ij}}{\partial \vartheta_j} & \dfrac{\partial^2 L}{\partial V_i \partial V_j}^0 + \delta_{lj} \dfrac{\partial P_{ij}}{\partial V_i} \dfrac{\partial P_{ij}}{\partial V_j} & \dfrac{\partial^2 L}{\partial V_i \partial \lambda_{pi}}^0 & \dfrac{\partial^2 L}{\partial V_i \partial \lambda_{q3}}^0 \\[6pt] \dfrac{\partial^2 L}{\partial \lambda_{pi} \partial \vartheta_j}^0 & \dfrac{\partial^2 L}{\partial \lambda_{pi} \partial V_j}^0 & 0 & 0 \\[6pt] \dfrac{\partial^2 L}{\partial \lambda_{q3} \partial \vartheta_j}^0 & \dfrac{\partial^2 L}{\partial \lambda_{q3} \partial V_j}^0 & 0 & 0 \end{bmatrix}$$

(97)

where $\delta_i = -\dfrac{\pi_{lli}}{s_{lli}} + \dfrac{\pi_{lui}}{s_{lui}}$

Gradient term modification:

$$t^T = \begin{bmatrix} \nabla_{P_1} L & \nabla_{P_2} L & \cdots & \nabla_{\theta_i} L & \nabla_{vi} L & \cdots & \cdots \end{bmatrix} \quad (98)$$

$$\tilde{t}^T = \begin{bmatrix} \nabla_{P_1} L + \delta_{g1} & \nabla_{P_2} L + \delta_{g2} & \cdots & \nabla_{\theta_i} L + \sum_{j=1}^{l} \delta_j \dfrac{\partial P_{ij}}{\partial \vartheta_i} & \nabla_{vi} L + \sum_{j=1}^{l} \delta_j \dfrac{\partial P_{ij}}{\partial V_i} & \cdots \end{bmatrix}$$

(99)

where $\delta_j = \dfrac{-1}{s_{llj}} + \dfrac{1}{s_{ulj}} \quad \delta_{gi} = \dfrac{-1}{s_{lpi}} + \dfrac{1}{s_{upi}} - (\pi_{lr} + \dfrac{\mu}{s_{lr}})$

9.4 Implementation Issues

9.4.1 INITIALIZATION

All interior point methods are sensitive to the initial estimate z^0 of the solution. An important feature of PDIP methods is that no initial feasible point is required. However, a start point must satisfy the positivity condition of slack variables and a barrier parameter $\mu^{(0)} > 0$ that causes the objective function logarithmic terms to dominate over the value of the original objective.

A flat start ($v = 1, \vartheta = 0$ for all the buses) can be adopted, which can guarantee the positivity of the slack variables. For other primal variables (P,Q, t etc.), the initial value can be determined by (l+u)/2, where l and u are there lower and upper limits respectively. Alternatively, a load flow program can be used to obtain the initial value of voltage and if any components are above their limits, two strategies are possible: (a) A correcting tool can be resorted to so that voltage violations are eliminated with a minimum number of control actions. (b) If the loss reduction goal is given preference and the violation is acceptable, the violation voltage bound can be initially broadened. Should this voltage lay within the limits during the iterative process, the relaxed bound would be reset to its original value, in any case, the optimization process would not deteriorate the initial state. Fig.3 compares flat start and converged power flow initializations. It shows that the latter can speed the convergence of the overall algorithm.

The slack variables are initialized according to

$$s_l = -h_l + h(z)^0$$
$$s_u = h_u - h(z)^0$$
(100)

The slack variables are positive for the given network state. In any case, if a slack variable is smaller than a certain threshold (e.g., 0.001), set that component to the threshold value. The dual variables corresponding to power flow equations, λ_p and λ_q, can be simply set to zero and the rest of the Lagrange multipliers can be evaluated from (89-90) for a selected μ. There are also many other sophisticated ways to produce such a starting point [13][18]. The best choice of a default starting point is still very much an open question.

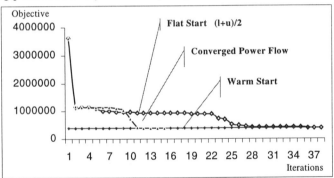

Fig.3 Effects of different initial points

9.4.2 CHOOSING STEP SIZE

The maximum step lengths are chosen to preserve the feasibility of all the problem variables. For the primal variables and dual variables, coefficients α_p, α_d are the primal and dual slack stepsizes, respectively and determined as follows

$$\alpha_p = \min\left\{\min_{\Delta s_{li}<0}\frac{-s_{li}}{\Delta s_{li}}, \min_{\Delta s_{ui}<0}\frac{-s_{ui}}{\Delta s_{ui}}, 1.0\right\} \tag{101}$$

$$\alpha_D = \min\left\{\min_{\Delta \pi_{li}<0}\frac{-\pi_{li}}{\Delta \pi_{li}}, \min_{\Delta \pi_{ui}>0}\frac{-\pi_{ui}}{\Delta \pi_{ui}}, 1.0\right\} \tag{102}$$

The separate step algorithm was first implemented by McShane, Monma, and Shanno[45] and has proven highly efficient in practice, significantly reducing the number of iterations to convergence. A new approximation to the optimal solution is computed from:

$$Z^{k+1} = Z^k + \alpha \cdot \alpha_p^k \cdot \Delta Z^k \tag{103}$$

$$S_u^{k+1} = S_u^k + \alpha \cdot \alpha_p^k \cdot \Delta S_u^k \tag{104}$$

$$S_l^{k+1} = S_l^k + \alpha \cdot \alpha_p^k \cdot \Delta S_l^k \tag{105}$$

$$\lambda_p^{k+1} = \lambda_p^k + \alpha \cdot \alpha_D^k \cdot \Delta \lambda_p^k \tag{106}$$

$$\lambda_q^{k+1} = \lambda_q^k + \alpha \cdot \alpha_D^k \cdot \Delta \lambda_q^k \tag{107}$$

$$\pi_l^{k+1} = \pi_l^k + \alpha \cdot \alpha_D^k \cdot \Delta \pi_l^k \tag{108}$$

$$\pi_u^{k+1} = \pi_u^k + \alpha \cdot \alpha_D^k \cdot \Delta \pi_u^k \tag{109}$$

where α is a step length parameter and $0.99 \le \alpha \le 0.9995$. K is the iteration step.

9.4.3 BARRIER FACTOR

A critical aspect of the proposed algorithm is the choice of barrier factor μ. It has great influence to the convergence of PDIP. It is usually estimated based on the predicted decrease of the duality gap. In linear programming, several schemes are proposed to chose μ. In refs. [19]~[22], the initial value is user defined and at intervening iterations, it is determined by:

$$\mu^{k+1} = \frac{-s_l^T \pi_l + s_u^T \pi_u}{2n\beta} \tag{110}$$

where $\beta > 1$, is a damping factor, n is the number of variables in the optimization problem. Fig 4 compares the influences of different choices of damping factors. Fig.5 gives the complementarily gaps with different start points.

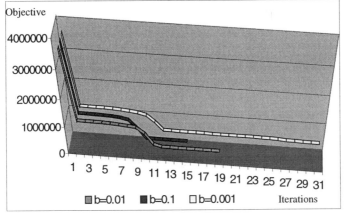

Fig.4 The influence of barrier parameter

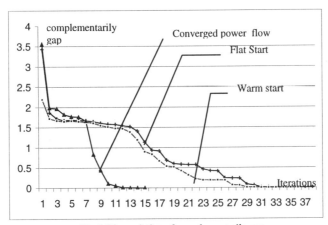

Fig.5. The variation of complementarily gap

9.4.4 Convergence criteria

For the LP problem, stopping criteria are usually defined in terms of the relative duality gap. For NLP, the algorithm terminates as both the duality gap and the mismatches of KKT are sufficient small. Ref. [37] points out, two or more of the following criteria should be satisfied to guarantee the convergence of PDIP.

• The duality-gap $D^k < \varepsilon_1$

- The maximum barrier factor $\mu^k < \varepsilon_2$
- The bus real/reactive power mismatch $\|g(x^k)\|_\infty < \varepsilon_3$
- $\|\Delta z\|_\infty < \varepsilon_4$
- The maximum Lagrangian gradient $\|\tilde{r}(x^k)\|_\infty < \varepsilon_5$

9.5 A Case Study on IPM Based Spot Pricing Algorithm

A practical, unambiguous and predictable method for active, reactive power and ancillary services real time pricing is provided in this section to demonstrate the application of IPMs, and two major concerns in the intriguing electricity market are included: how to update the existing OPF to a pricing tool and how to avoid the "go" "no-go" gauge. IEEE 30-bus system case study is reported

9.5.1 SPOT PRICES AND THEIR DECOMPOSITION

Spot pricing for active and reactive power can be defined as:

$$\rho_{pi} = \frac{\partial C}{\partial P_i} \tag{111}$$

$$\rho_{qi} = \frac{\partial C}{\partial Q_i} \tag{112}$$

where C is the total system production cost. ρ_{pi} and ρ_{qi} are spot prices of real power and reactive respectively. P_i, Q_i are active and reactive power injections of bus i This problem can be summarized as a optimization problem to minimization system production cost limited by system equality and inequality constraints. In this paper, C includes both active and reactive power production costs which are simulated as quadratic curves.

$$\text{Min } F = \sum_{i=1}^{m}(f_{pi}(P_{gi}) + f_{qi}(Q_{gi})) \tag{113}$$

Where $f_{pi}(P_{gi}) = CP_{gi}^2 + BP_{gi} + A$, $f_{qi}(Q_{gi}) = C_r Q_{gi}^2 + B_r Q_{gi} + A_r$

The constraints taken into account in this section include: power flow balance, voltage limits, generator output limits, line flow limits and tap changing limits etc. Mathematically, the optimal spot price problem can be written as:

$$\begin{aligned} \min \quad & F(z) \\ \text{s.t.} \quad & g(z) = 0 \\ & h_l \le h(z) \le h_u \end{aligned} \tag{114}$$

From economic theory, the Lagrangian multipliers corresponding to active and reactive power balance, λ_{pi} and λ_{qi}, have the same economic meanings as SRMC (Short Run Marginal Cost) of active and reactive power respectively [28][29]:
The spot prices of node i:

$$\rho_{pi} = \lambda_{pi} \qquad (115)$$

$$\rho_{qi} = \lambda_{qi} \qquad (116)$$

and we can further decomposed them into different price components corresponding to the original concepts of spot price provided in ref.[27]:

$$\lambda_{pi} = \left(1 - \frac{\partial P_L}{\partial P_i}\right)\frac{\partial F}{\partial P_s} - \frac{\partial Q_L}{\partial P_i}\frac{\partial F}{\partial Q_s} - \sum_{j \notin h_o} \frac{\partial h_j}{\partial P_i}(\pi_{lj} + \pi_{uj}) \qquad (117)$$

$$\lambda_{qi} = -\frac{\partial P_L}{\partial Q_i}\frac{\partial F}{\partial P_s} + \frac{\partial F}{\partial Q_s}(1 - \frac{\partial Q_L}{\partial Q_i}) - \sum_{j \notin h_o} \frac{\partial h_j}{\partial Q_i}(\pi_{lj} + \pi_{uj}) \qquad (118)$$

Both λ_{pi} and λ_{qi} consist of three components: the first part is active power loss compensation price, the second part corresponds to reactive power loss and the third term is defined as security price. $\frac{\partial F}{\partial P_s}$ is also called "system lambda" in economic dispatching. Similarly, $\frac{\partial F}{\partial Q_s}$ represents "reactive power system lambda", denoted as $\lambda_{ps}, \lambda_{qs}$ respectively. Therefore, the Lagrangian multipliers obtained from OPF can be used to evaluate the spot prices of active and reactive power directly, with extra sensitivities calculation, we can obtain every component of spot price in (117) and (118).

9.5.2 ALGORITHM

Step 0: Market options and system operation conditions input.
Step 1: Solve initial power flow or use flat start power flow to a choose starting point that strictly satisfies the positivity conditions.
Step 2: Form the Newton system, modify Hessian matrix & right hand vector.
Step 3: Compute Newton directions by solving the Newton system.
Step 4: Determine iteration step lengths for both primal and dual variables.
Step 5: Test convergence. If convergence criteria are satisfied, go to step 7, otherwise go to step 6.
Step 6: Update barrier factor and go back to step (2).
Step 7: Calculate sensitivities matrix, decompose spot price into appropriate parts and output results.

9.5.3 NUMERICAL RESULTS

The IEEE 30-bus system has six generators and four ULTC transformers. A loading condition of 324.29MW and 114.19MVAR is used to perform the proposed method. The units of all quantities are given in per unit. The quadratic cost curves are showed in appendix B. All bus voltage upper limits are 1.05, lower limits are 0.95. The spot prices of active and reactive power are shown in (a) and (b) respectively in every case. The unit of spot price for active power and reactive power are ¢/MWh and ¢/Mvarh. Define:

$$lp_i^* = \lambda_{ps}(1 - \frac{\partial P_L}{\partial P_i}) \qquad lq_i^* = \lambda_{qs}(1 - \frac{\partial Q_L}{\partial Q_i})$$

$$lp_i^{**} = lp_i^* - \lambda_{qs}\frac{\partial Q_L}{\partial P_i} \qquad lq_i^{**} = lq_i^* - \lambda_{ps}(\frac{\partial P_L}{\partial Q_i})$$

$$lp_i^{***} = lp_i^{**} + \sum_j (\pi_{llj} + \pi_{ulj})\frac{\partial P_{lj}}{\partial P_i} + \pi_r \frac{\partial R}{\partial P_i}$$

$$lq_i^{***} = lq_i^{**} + \sum_j (\pi_{llj} + \pi_{ulj})\frac{\partial P_{lj}}{\partial Q_i} + \sum_j (\pi_{lvj} + \pi_{uvj})\frac{\partial V_j}{\partial Q_i}$$

The results obtained from the different operations are presented below, notice that because lp_i^{***} and lq_i^{***} are too close to the λ_{pi} and λ_{qi} to be shown in the following figures.

1. Normal operation conditions

Fig 6(a) reveals that active power spot price is dominated by system lambda and network loss compensation term in normal operation conditions. The congestion and security terms are relatively small. This conclusion is the same as [29]. Reactive power spot price at all buses are dominated by reactive power generation cost and real power loss compensation item, as shown in Fig 6(b). Comparing with active power, reactive power spot price is much more volatile. The security prices for both active and reactive power are negligible.

2. Congestion conditions

Line constraints force the use of a higher loss path to satisfy the demand requirements, and may also require the reallocation of generation which would increase the total cost. Figs 7(a) and 7(b) show the variations of active power and reactive power spot prices when the power flow limits for line 3 are approached

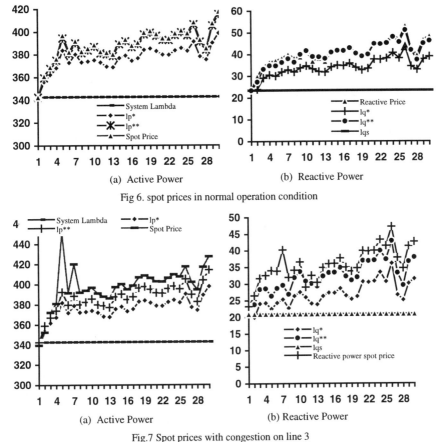

Fig 6. spot prices in normal operation condition

Fig.7 Spot prices with congestion on line 3

(The power flow limits is 68 MW, the optimisation result is 67.999 MW). Security prices are increased for all buses, especially for Buses 5 and 7, for they are most sensitive to the variation of power flow on line 3. From fig 7(b), transmission limits affects reactive power spot price much in the same way as it does real power price.

3. Voltage violation

Voltage profiles are determined mainly by reactive power supply. Therefore, voltage constraints as well as the generation and consumption pattern of reactive power by the customers have the greatest impact on the spot prices of reactive power. Tightening the voltage lower limits of bus 7 (increase it from 0.95 to 0.99), the reactive power security prices are increased rapidly, because they are dominated by voltage constraints in this situation. The phenomenon can be interpreted by equation (118). To prevent the reactive spot price form increasing too much, we provide new reactive sources by adding reactive power compensation (shunt capacitance) at bus 7. The spot prices drop largely, and security prices go back to the normal level. Notice this act changes the active power loss price components as well.

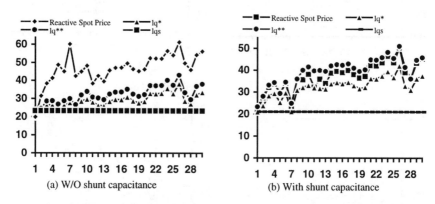

(a) W/O shunt capacitance   (b) With shunt capacitance

Fig. 8 Reactive power spot prices with bus 3 voltage violates limits

(4) Alleviating "go" "no go" gauge

In order to illustrate how IOSP alleviates "go" "no go" gauge, fig. 9 shows the response of active power spot price of bus 5 due to active power flow limits on line 3. Under normal operation conditions, the line flow limit of line 3 is set to 100MW and the active power flow on line 3 is 72MW. When the line flow limit is decreasing from 100MW to 74MW, the spot price of bus 5 keeps almost the same value. This is because the security price is negligible. Once it is approaching to 72MW, the security price of bus 5 starts to increase gradually. The closer to the boundary (shown in dark black in Figs 9 and 10) the limit, the bigger the spot

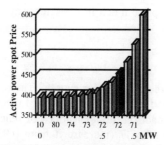

Fig. 9 Active power spot price at bus 5
vs line flow limit of line 3

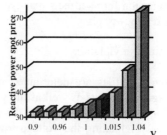

Fig.10 Reactive power spot price
vs voltage lower limit at bus 5

price. After the limit is set to be smaller than 72MW, the spot price increases even more rapidly. For example, when the limit is decreased to 71MW, the spot price of bus 5 increases to the point nearly 2 times of the normal operation price. Fig.10 shows the response of reactive power spot price of bus 5 due to the voltage lower limit of bus 5. Similar phenomenon can be observed. PDIP ensures the security price changes exponentially, which makes the "go" "no-go" gauge avoidable.

9.6 Conclusions

This chapter deals with Interior Point methods. SCED and OPF are selected to outline its applications in power systems. Special emphasis is given to PDIP method which has attracted most of power system optimization researchers. Many important implementation issues are discussed and the following points have been noted

- Converged power flow starting point accelerates the convergence.
- Reduced KKT System enhances the performance
- Super Sparsity techniques can be applied
- Barrier factor plays an important role in the process of convergence.
- PCPDIP performs better than pure PDIP
- Due to the introduction of soft constraints, highly volatile spot prices can be avoided.

The authors believe not only their efficiency to solve LP and NLP problems, but also their capability to combine with Newton OPF, will make IPMs very successful in power system optimization.

9.7 Acknowledgment

The authors gratefully acknowledge the support of the NGC and the useful discussions with Prof M R Irving of Brunel, Prof J F Macqueen and Dr Pals of the NGC.

9.8 References

1. K.R. Frisch, 1955, The Logarithmic potential method for convex programming, Institute of Economics, University of Oslo, Oslo, Norway (unpublished manuscript)
2. A.V. Fiacco and G.P .McComick, Non-linear Programming: Sequential Unconstrained Minimisation Techniques , *John Wiley & Sons*, New York, 1968.
3. I.I. Dikin, Iterative Solution of Problems of Linear and Quadratic Programming, *Doklady Akademii Nauk SSSR* 174:747-748,1967. Translated into English in Soviet Mathematics Doklady, 8: 674-675.

4. N.K. Karmarkar, 1984. A New Polynomial-Time Algorithm for Linear Programming, *Combinatoria 4*, 373-395
5. I.J. Lusting, R.E. Marsten, D.F. Shanno, Interior Point Methods for Linear Programming: Computation State of the Art, *ORSA Journal on Computing*, Vol.6, No.1. Winter 1994
6. G. Strang, Linear Algebra and Its Applications, 405-411, *Harcout Brace Jovanovich*, 1988
7. J. Gondzio, T. Terlaky, Advanced in Linear Programming, 1996
8. R.J. Vanderbei, Interior Point Methods: Algorithms and Formulations, *ORSA Journal on Computing*, Vol.6, No.1 Winter 1994
9. K.A. Mcshane, C.L. Monma and D.F. Shanno, 1989, An Implementation of Primary Dual Interior Point Method For Linear Programming, *ORSA Journal on Computing* 70-83
10. K.A. Clements, P.W. Davis, K.D. Prey, "An Interior Point Algorithm for Weighted Least Absolute Value Power System State Estimation", *IEEE paper 91 WM 235-2 PWRS*.
11. G. Irisarri, L.M. Kimball, K.A. Clements, A. Bagchi and P.W. Davis, "Economic Dispatch with Network and Ramping Constraints via Interior Point Methods", *IEEE Trans. PWRS*, Vo.13, No.1, February 1998
12. L.S. Vargas, V.H. Quintana and A. Vannelli, "A Tutorial Description of An Interior Point Method and Its Applications to Security Constrained Economic Dispatching", *IEEE Trans. PWRS*, Vo.8, No.3, August 1993
13. X. Yan, V.H. Quintana, "An Efficient Predictor-Corrector Interior Point Algorithm for Security-Constrained Economic Dispatching", *IEEE Trans. PWRS*, Vo.12, No.2, February 1997
14. S. Granville, "Optimal Reactive Dispatch Through Interior Point Methods", *IEEE Trans. PWRS*, Vo.9, No.1, February 1994
15. J.A. Momoh, S.X. Guo, E.C. Ogbuoriri and R. Adapa, " The Quadratic Interior Point Method Solving Power System Optimisation Problems", *IEEE Trans. PWRS*, Vo.9, No.3, August 1994
16. G.R. Mda Costa, "Optimal Reactive Dispatch Through Primal-Dual method", *IEEE Trans. PWRS*, Vo.12, No.2, May 1994
17. J.L Martinez Ramos, A. Gomez Exposito, V.H. Quintana, " Reactive Power Optimisation by Interior Point Methods: Implementation Issues", *12th Power Systems Computation Conference*, Dresden, August 19-23, 199
18. Y.C. Wu, A.S. Debs and R.E. Marsten, "A Direct Nonlinear Prdictor-Corrector Primal-Dual Interior Point Algorithm for Optimal Power Flow", *IEEE Trans. PWRS*, Vo.9, No.2, May 1994
19. S. Granville, J.C.O. Mello, A.C.G. Melo, " Application of Interior Point Methods to Power Flow Unsolvability", *IEEE Trans. PWRS*, Vo.11, No.2, May 1996
20. J.C.O. Mello, A.C.G. Mello, S. Granville, "Simultaneous Transfer Capacity Assessment by Combining Interior Point Methods And Monte Carlo Simulation", *IEEE Summer Meeting*, 1996

21. G.D. Irisarri, X. Wang, J. Tong, S. Mokhari, "Maximum Loadability of Power Systems using Interior Point Non-Linear Optimisation Method", *IEEE Trans. PWRS*, Vo.12, No.1, February 1997
22. K. Xie, Y.H. Song, G.Y. Liu and E.K. Yu, "Real Time Pricing of Electricity Using Interior Point Optimal Power Flow Alogorithm",*UPEC'98*, Edinburgh, UK, 1998
23. C.N. Lu, MR. Unum, "Network Constrained Security Control Using Interior Point Algorithm", *IEEE Trans. PWRS*, Vo.8, No.3, August 1993
24. D.I. Sun, B. Ashley, B. Brewer, A. Hughes, W.F. Tinney, "Optimal Power Flow by Newton Approach", *IEEE Transactions on Power Systems*, Vol. PAS-103, No.10, October 1984
25. Y.G. Hao, "Study of Optimal Power Flow and Short-term Reactive Power Scheduling", *PhD Thesis*, Electric Power Research Institute (China), 1997
26. K.A. Mcshane, C.L. Monma and D.F. Shanno, 1989, An Implementation of a Primal-Dual Interior Point for Linear Programming, *ORSA Journal on Computing* 1, 70-83.
27. F. Schweppe, M. Caramanis, R. Tabors, and R. Bohn, Spot Pricing of Electricity, *Kluwer Academic Publishers*, Boston, MA, 1988
28. M.L. Baughman, S.N. Siddiqi, "Real Time Pricing of Reactive Power: Theory and Case Study Results", *IEEE Transactions on Power Systems*, Vol. 6, No. 1, February 1993
29. A.A. EL-Keib, X. Ma, "Calculating of Short-Run Marginal Costs of Active and Reactive Power Production", *IEEE Transactions on Power Systems*, Vol.12, No. 1, May 1997
30. J.D. Finney, H.A. Othman, W.L. Rutz, Evaluating Transmission Congestion Constraints in System Planning, *IEEE Transactions on Power Systems*, Vol. 12, No. 3, August 1997
31. I. Adler, N.K. Karmarkar, M.G.C. Resende and G. Veiga, 1989, An Implementation of Karmarkar's Algorithm for Linear Programming, *Mathematical Programming* 44, 297-335
32. R.D.C. Monteiro and I. Adler,1989, Interior Path Following Primal-Dual Algorithms: Part 1: Linear Programming, *Mathematical Programming* 44, 27-41
33. P.E. Gill, W. Murray, M.A. Saunders, J.A. Tomlin and MH. Wright, 1986. On Projected Newton Barrier Method for Linear Programming and an Equivalence to Karmarkar's Projective Method, *Mathematical Programming* 36, 409-429
34. N. Megiddo, 1989, Pathways to Optimal Set in Linear Programming, pp.131-138 in Progress in Mathematical Programming,: Interior Point and Related Methods, N.Megiddo (ed.), *Springer Verlag*, NY
35. S. Mehrotra, 1992, On the Implementation of a Primal-Dual Interior Point Method, *SIAM Journal on Optimisation* 2:4, 575-601

36. A. Canizares, F.L. Alvarado, C.L. DeMarco, I. Dobson, W.F. Long, " Point of Collapse Methods Applied to AC/DC Power Systems", *IEEE Transactions on Power Systems*, Vol. 7, No. 2, August 1992
37. V.H. Quitana, A. Gomez and J.L. Martinez, "Nonlinear Optimal Power Flow by a Logrithmic-Barrier Primal-Dual Algorithm", *Proc. Of NAPS*, 1995
38. K.A. Clements, P.W. Davis and K.D. Frey, "Treatment of Inequality Constraints in Power System State Esitimation", *IEEE Transactions on Power Systems*, Vol. 10, No. 2, 567-573, May 1995
39. J. Carpentier, "Contribution to the Economic Dispatch problems", *Bull. Soc. France Elect*, Vol.8, 431-437, August, 1962
40. R. J. Vanderbei, D.F. Shanno, " An Interior Point Algorithm for Nonconvex Nonlinear Programming, " *1991 Mathematics Subject Classification Primary 90C30*, Secondary 49M37, 65K05
41. M. Huneault, F.D. Galliana, "A Survey of the Optimal Power Flow Literature," *IEEE Transactions on Power Systems*, Vol. 6, No. 2, 762-770, May 1991
42. H. Wei, H. Sasaki, J. kubokawa, R. Yokoyama, " An Interior Point Nonlinear Programming for Optimal Power Flow Problems with a Novel Data Structure", *IEEE Transactions on Power Systems*, Vol. 13, No. 3, 870-877, August 1998
43. H. Wei, H. Sasaki, J. Kubokawa, R. Yokoyama, " An Interior Point Nonlinear Programming for Power System Weighted Nonlinear L_1 Norm State Estimation", *IEEE Transactions on Power Systems*, Vol. 13, No. 2, 870-877, May 1998
44. X. Wang, G.C. Ejebe, J. Tong, J.G. Waight, "Preventive/Corrective Control for Voltage Stability Using Direct Interior Point Method", *IEEE Trans on PWRS*, Vol. 13, No. 3, August 1998
45. K.A. Mcshane, C.L. Monma, D.F. Shanno, An Implementation of a Primal-Dual Interior Point Methods For Linear Programming, *ORSA Journal on Computing* 1, 70-83
46. H. Wei, H. Sasaki, J. kubokawa, " A Decoupled Solution of Hydro-thermal Optimal Power Flow by Means of Interior Point Method and Network Programming", *IEEE Transactions on Power Systems*, Vol. 13, No. 2, , May 199
47. K. Ponnambalam, V.H. Quintana, A. Vannelli, " A Fast Algorithm for Power System Optimization Problems Using An Interior Point Method", *IEEE Transactions on Power Systems*, Vol. 7, No. 2, May 1992

Appendix 1

Table 1 Economic Parameters of Generators

ID	Bus	A	B	C	Ar	Br	Cr
1	1	100	2	0.0037	10	0.2	0.00037
2	2	100	1.75	0.0175	10	0.17	0.00175
3	5	100	1	0.0625	10	0.1	0.00625
4	8	100	3.25	0.0083	10	0.32	0.00083
5	11	100	3	0.025	10	0.3	0.0025
6	13	100	3	0.025	10	0.3	0.0025

Chapter 10

ANT COLONY SEARCH, ADVANCED ENGINEERED-CONDITIONING GENETIC ALGORITHMS AND FUZZY LOGIC CONTROLLED GENETIC ALGORITHMS: ECONOMIC DISPATCH PROBLEMS

Y.H. Song, C.S.V. Chou, I.K. Yu And G.S. Wang
Department of Electrical Engineering and Electronics
Brunel University, Uxbridge
UB8 3PH, UK

10.1 Introduction

Various novel search algorithms have been proposed in recent years, some of which have been described in some details in the previous chapters. This chapter is to present some newly developed techniques in the areas of natural phenomena inspired algorithms and hybridisations. The three algorithms described are: ant colony search [1,2], advanced engineered-conditioning genetic algorithm [3] and fuzzy logic controlled genetic algorithms [4]. Some test results on economic dispatch problems have also been given.

10.2 Ant Colony Search Algorithms

For the last few years there has been a growing interest in algorithms inspired by the observation of natural phenomena to help solve complex computational problems. In this section, a novel co-operative agent algorithm, Artificial Ant Colony Search Algorithm (ACS), is introduced. The ACS was inspired by the observation of the behaviour of ant colonies. Ant Colony Search Algorithms [1] have recently been proposed as powerful tools to solve some order based problems such as travelling salesman problem (TSP) and quadratic assignment problem.

10.2.1 THE BEHAVIOUR OF REAL ANTS

Ant colony search algorithms, to some extent, mimic the behaviour of real ants. It is interesting to understand how ants, which are almost blind animals with very simple individual capacities acting together in a colony, can find the shortest route between the ant's nest and a source of food. As is well known, real ants are capable of finding the shortest path from food sources to the nest without using visual cues. They are also capable of adapting to changes in the environment, for example, finding a new shortest path once the old one is no longer feasible due to a new obstacle. The studies by ethnologists reveal that such capabilities ants have are essentially due to what is called "pheromone trails" which ants use to communicate information among individuals regarding path and to decide where to go. Ants deposit a certain amount of pheromone while walking, and each ant probabilistically prefers to follow a direction rich in pheromone rather than a poorer one.

The process can be clearly illustrated by Figure 1. In Figure 1a ants are moving on a straight line which connects a food source to the nest. Once an obstacle appears as shown in Figure 1b, the path is cut off. Those ants which are just in front of the obstacle cannot continue to follow the pheromone trail and therefore it can be expected that they have the same probability to turn right or left. Figure 1c, these ants which choose by chance the shorter path around the obstacle will more rapidly reconstitute the interrupted pheromone trail compared to those which choose the longer path. Hence, the shorter path will receive a higher amount of pheromone in the time unit and this will in turn cause a higher number of ants to choose the shorter path. Due to this positive feedback (autocatalytic) process, very soon all ants will choose the shorter path.

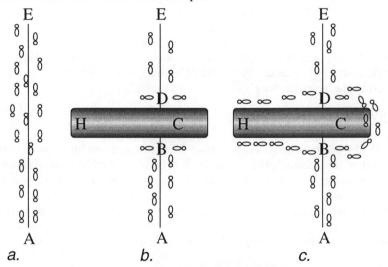

Figure.1 Behaviour of Real Ants

10.2.2 ANT COLONY SEARCH ALGORITHMS

The structure of a simple Ant Colony Search Algorithm is shown in Figure 2.

(1) Initialize A(t): The problem parameters are encoded as a real number. Before each run, the initial population (Nest) of the colony are generated randomly within the feasible region which will crawl to different directions at a radius not greater than R.

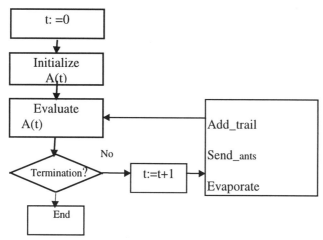

Figure 2 Structure of simple ACSA

(2) Evaluate A(t): The fitness of all ants are evaluated based on their objective function.

(3) Add_trail: Trail quantity is added to the particular directions the ants have selected in proportion to the ants' fitness.

(4) Send_ants A(t): According to the objective function, their performance will be weighed as fitness value which directs influence to the level of trail quantity adding to the particular directions the ants have selected. Each ant chooses the next node to move taking into account two parameters: the visibility of the node and the trail intensity of trail previously laid by other ants. The send_ants process sends ants by selecting directions using Tournament selection based on the two parameters. The k-th ant starting from node i decides to move to node j on the basis of probability p_{ij}^k defined as follows:

$$p_{ij}^k(t) = \begin{cases} \dfrac{[\tau_{ij}(t)]^\alpha \cdot [\eta_{ij}]^\beta}{\sum_{k \in allowed_k}[\tau_{ik}]^\alpha \cdot [\eta_{ik}]^\beta} & \text{if } j \in allowed_k \\ 0 & \text{otherwise} \end{cases} \qquad (1)$$

Where:

Visibility, $\eta_{ij} = |\mu - \Delta F|$ \qquad (2)

Move value, ΔF = original total cost - new total cost

μ, α, β are the heuristically defined parameters.

Intensity trail, $\tau_{ij}(t)$ on edge (i,j) at time t. Each ant at time t chooses the next node, where it will be at time t+1. For 1 iteration of ant colony search algorithm, the m moves carried out by the m ants in the interval (t, t+1), then every n iterations of the algorithm each ant has completed a tour. At this point the trail intensity is updated as:

$$\tau_{ij}(t+n) = \rho \cdot \tau_{ij}(t) + \Delta \tau_{ij} \qquad (3)$$

where: ρ is a coefficient of persistence of the trail during a cycle which is heuristically defined.

$$\Delta \tau_{ij} = \sum_{k=1}^{m} \Delta \tau_{ij}^{k} \qquad (4)$$

where: $\Delta \tau_{ij}^{k}$ is the quantity per unit length of trail substance laid on edge (i,j) by the k-th ant between time t and t+n.

(5) Evaporate: Finally, the pheromone trail secreted by an ant eventually evaporates and the starting point (nest) is updated with the best tour found.

10.2.3 THE CHARACTERISTICS OF ACS

There are some attractive properties of ant colony search algorithm when compared with other methods:

(i) Distributed Computation - Avoid Premature Convergence:

Conventionally, a top-down approach has been adopted. Scientists choose to work on system simplified to a minimum number of components in order to observe "pure" effects. Ant colony search algorithm often simplifies as much as possible the components of the system, for a purpose to take into account their large number and bottom-up approach. The power of the massive parallel fashion in ACSA is able to cope with incorrect, ambiguous or distorted information which are often found in nature. The computational model contains the dynamics which is determined by the nature of the local interactions between the many elements (artificial ants).

(ii) Positive Feedback - Rapid Discovery of Good Solution:

The unique inter-ant communication involves a mutual information sharing while solving a problem. Occasionally, the information exchanged may contain errors and should alter the behaviour of the ants receiving it. As the search proceeds, the new population of ants often containing states of higher fitness will affect the search behaviour of the others and will eventually gain control over the other agents while at the same time actively exploiting inter-ant communication by mean of the pheromone trail laid on the path. The artificial ant foraging behaviour dynamically reduces the prior uncertainty about the problem at hand. As ants doing a task can also be either "successful" or "unsuccessful" and can switch between these two according to how well the task is performed. Unsuccessful ants also have a certain chance to switch to be inactive, and successful ants had a certain chance to recruit inactive ants to their task. Therefore, the emerging collective effect is a form of autocatalytic behaviour, in that more ants following a particular path, the more attractive this path becomes for the next ants which should meet it. It can give rise to global behaviour in the colony.

(iii) Use of Constructive Greedy Heuristic - Find Acceptable Solutions In The Early Stage of The Process:

Based on the available information collected from the path (pheromone trail level and visibility), the decision is made at each step as a constructive way by the artificial ants, even if each ant's decision always remains probabilistic. It tends to evolve a group of initial poorly generated solution to a set of acceptable solutions through successive generations. It uses objective function to guide the search only, and does not need any other auxiliary knowledge. This greatly reduces the complexity of the problem. The user only has to define the objective function and the infeasible regions (or obstacle on the path).

10.3 Engineered-Conditioning GAs

Computational efficiency and reliability are the major concerns in the application of genetic algorithms (GAs) to practical problems. Effort has been made in two directions to improve the performance of GAs: the investigation of advanced genetic operators and the development of genetic algorithm hybrids. In this section, an Advanced Engineered-Conditioning Genetic Algorithm Hybrid (AEC-GA) is briefly described, which is a combination strategy involving local search algorithm and genetic algorithm.

10.3.1 ENGINEERED-CONDITIONING OPERATOR

Engineered-conditioning operator based on the first-order gradient method can be seen as a DNA repair system found in natural genetic mechanisms [5]. It is visualized that GA-hybrid exhibits more reliable than the pure GA as they compliment each other. The local search algorithm will try to optimize locally, while the genetic algorithm will try to optimize globally. The EC-GA hybrid has been used in the past to achieve higher reliability in solving MFD problems [6]. Miller, Potter, Gandham and Lapena (1993) used three local search or "conditioning" tests in the ECO. These are, in order of application, the superset, substitute, and subset evaluation tests. In term of the running time, the complexity of this test is $O(n^2)$. In order to overcome the time-consuming trial and error process, this ECO is incorporated with problem specific knowledge to ensure that it works well with GA on power economic dispatch. The new ECO proposed is more accurate and the complexity of the test is only $O(n)$.

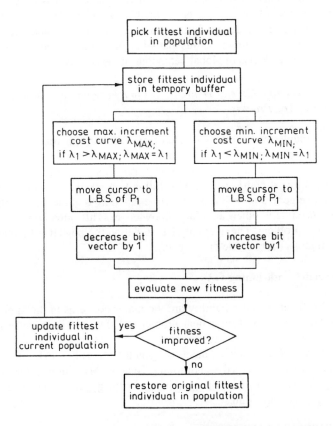

Figure 3 Engineered-conditioning operator

The engineered-conditioning operator for power economic dispatch works as follow: In the end of the run of each generation, the fittest individual is copied into a temporary population waiting for the next process. The ECO will try to compare each incremental cost curve (X) of this individual and select a particular pair of units to move. In this case two units with extreme X are selected. It would be possible to move the power of these two units by flipping the bit on and off such that the average incremental costs will be minimized while maintaining no power balance error on the system. The individual will introduce back into the current population if any further improvement can be made. The implementation of engineered-conditioning operator is shown in Figure 3.

10.3.2 ADVANCED ENGINEERED-CONDITIONING GAS

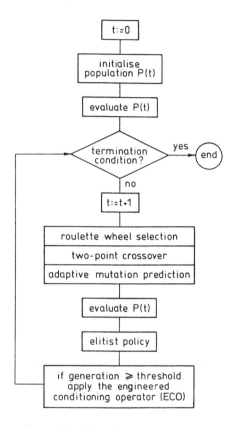

Figure 4 Flowchart of proposed AEC-GAs

By coupling the advanced genetic operators with engineered-conditioning operator, Advanced Engineered-Conditioning Genetic Algorithms (AEC-GAs) is proposed. The local improvement offered by the ECO is used to improve the diagnosis obtained from the global searching of the genetic algorithm. The AEC-GA hybrid works as follows: run the genetic algorithm, but after each generation, removes the fittest individual, apply the ECO to this individual, and introduce the hopefully improved individual back into the population. The hybridization techniques are very much problem specific and care should be taken when enhancing genetic algorithms so as not to harm the benefits offered by genetic algorithms. Since the overuse of local improvement operators in the early stages of GA evolution can destroy the randomness of the population. This early reduction in diversity can possible lead to premature convergence and result in overall poor performance. Therefore, a phased-in AEC-GA has been adopted to overcome this pitfall. The ECO is applied only when the following condition is satisfied: Generation>Threshold. The flow chart of the ACE-GA is illustrated in Figure 4.

10.4 Fuzzy Logic Controlled Genetic Algorithms

Although this basic GA has been applied to some problems, its drawbacks prevent the acceptance of the theoretic performances claimed. Thus various techniques have been studied to improve genetic search. These include: (1) using advanced string coding; (2) generating an initial population with some prior knowledge; (3) establishing some better evaluation function; (4) properly choosing parameters and (5) using advanced genetic operators. Particularly, as genetic algorithms are distinguished from others by the emphasis on crossover and mutation, more recently much attention and effort has been devoted to improving them. In this respect, two-point, multi-point and uniform crossover, and variable mutation rate have been recently proposed. In this section, more advanced genetic operators have been presented which are based on fuzzy logic with the ability to adaptively/dynamically adjust the crossover and mutation during the evolution process.

Fig 5 presents the block diagram of a fuzzy-controlled genetic algorithm, in which two on-line fuzzy logic controllers are used to adapt the crossover and mutation. The objective here is to provide a significant improvement in the rate of convergence. The fuzzy controller in Fig 5 consists of four principal components: (1) fuzzification interface, which converts crisp input data into suitable linguistic values; (2) Fuzzy rule base, which consists of a set of linguistic control rules incorporating heuristics that are used for the purpose of achieving a faster rate of convergence; (3) Fuzzy inference engine, which is a decision-making logic that employs rules from the fuzzy rule base to infer fuzzy control actions in response to fuzzied inputs; and (4) Defuzzification interface, which yields a crisp control action from an inferred fuzzy control action. In the rest of the section, detailed description of the design of fuzzy crossover and fuzzy mutation controllers will be given.

10.4.1 FUZZY CROSSOVER CONTROLLER

The fuzzy crossover controller is implemented to automatically adjust the crossover probability during the optimisation process. The heuristic updating principles of the crossover probability is if the change in average fitness of the populations is greater than zero and keeps the same sign in consecutive generations, then the crossover probability should be increased. Otherwise the crossover probability should be decreased.

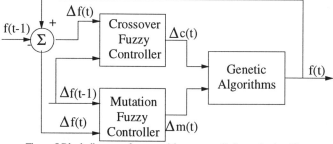

Figure 5 Block diagram of proposed fuzzy controlled genetic algorithm

10.4.2 FUZZY MUTATION CONTROLLER

The mutation operation is determined by the flip function with mutation probability rate, and the mutated bit is randomly performed. The mutation probability rate is automatically modified during the optimisation process based on a fuzzy logic controller. The heuristic information for adjusting the mutation probability rate is if the change in average fitness is very small in consecutive generations, then the mutation probability rate should be increased until the average fitness begins to increase in consecutive generations. If the average fitness decreases, the mutation probability rate should be decreased.

The inputs to the mutation fuzzy controller are the same as those of the crossover fuzzy controller, and the output of which is the change in mutation $\Delta m(t)$. The design of the membership function, decision and action tables for the fuzzy mutation controller is similar to these for the fuzzy crossover controller.

10.5 Test Results On Economic Dispatch Problems

10.5.1 THE SIX-UNIT ECONOMIC DISPATCH PROBLEM

The first test is carried out on pure economic dispatch of a six generator system. The data employed in this paper are obtained from [7]. The fuel cost function of each unit is a quadratic function of the generator real power output, and the output limits are given as follows:

$F_1 = 0.001562 P_1^2 + 7.92 P_1 + 561.0 \quad 100 < P_1 < 600$
$F_2 = 0.00194 P_2^2 + 7.85 P_2 + 310.0 \quad 100 < P_2 < 400$
$F_3 = 0.00482 P_3^2 + 7.97 P_3 + 78.0 \quad 50 < P_3 < 200$
$F_4 = 0.00139 P_4^2 + 7.06 P_4 + 500.0 \quad 140 < P_4 < 590$
$F_5 = 0.00184 P_5^2 + 7.46 P_5 + 295.0 \quad 110 < P_5 < 440$
$F_6 = 0.00184 P_6^2 + 7.46 P_6 + 295.0 \quad 110 < P_6 < 440$

The load demand is assumed to be 1800MW. For simplicity, transmission losses are ignored in the test.

10.5.2 RESULTS BY ACS, AEC-GAS AND FCGAS

In the ACS, the following parameters have been chosen heuristically: Number of ants = 100, Number of cycle = 20, α = 0.5, β = 0.05, μ = 10, ρ = 0.5, Q = 50. The results with load demand of 1800MW are compared with conventional genetic algorithm (CGA) with a similar condition. (i.e. 100 individuals and 20 generations).

In the AEC-GA, The parameters used in AEC-GA and CGA are as follows: Population size = 100; Sub-chromosome length = 11; Chromosome length = 66; Optimized parameter = 6; Crossover rate = 0.98; Initial mutation rate = 0.01. In reality, the load demand will be unpredictable, and it varies throughout the day.

The parameters used in conventional GA and FCGA are: Population size=100; Sub-chromosome length (i.e. for each unit)=10; Optimised parameters=6; Chromosome length=60; Initial crossover rate=0.5; Initial mutation rate=0.01; Desired generations=100.

Table 1 Comparison between CGA, ACSA, AEG and FCGA

	Unit 1	Unit 2	Unit 3	Unit 4	Unit 5	Unit 6	Computation Time (S)	Fuel Cost
Newton	184.00	166.20	54.40	590.00	402.70	402.70		16609.57
CGA	222.42	190.73	95.36	555.63	367.92	367.92	19.66	16589.05
ACSA	248.27	217.36	74.94	588.37	335.78	335.28	23.4	16579.33
AEC-GA	248.07	217.73	75.3	587.7	335.6	335.6	7.1	16579.33
FCGA	250.49	215.43	109.9	572.84	325.66	325.66	10.44	16585.85

The fuel costs and the unit dispatch in Table 1 for three method (ACS, AEC-GA and FCGA) clearly shows the performance improvement over the CGA. Having more ants in the colony which searches at the same time results in higher performance as a larger space is covered. Moreover, the performance increases in a nonlinear manner. All the ants seems to have a better vision to cross over the false peaks, thus dramatically identify the solution cluster. This indicates that there are constructive interactions taking place, resulting in an emergent collective computational ability that is higher than the sum of individual. For clarity, Figure 6 illustrates the behaviours of only 3 ants out of 100. To appreciate the evolutionary process, Figures 7 illustrates the optimization progress during each run for both on-line and off-line performance of AEC-GA with load demand of 1800MW. Figures 8 (a) and (b) respectively show the cost curves of FCGAs and CGAs when load demand is assumed to be 1800MW during the optimisation process.

The cost curve of Figure 8(b) is oscillating from one generation to another. This is mainly because the crossover rate and mutation rate have been kept constant in CGAs. This causes problems to set criteria to stop the search for an optimal solution. On the other hand, the cost curve of the proposed FCGA indicates the solution is improved during each generation.

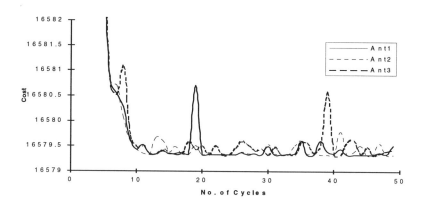

Figure 6 The evolutionary process of ants within a colony [for simplicity, only 3 ants are shown]

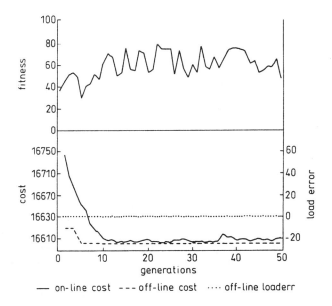

Figure 7 Evolutionary process of AEC-GA

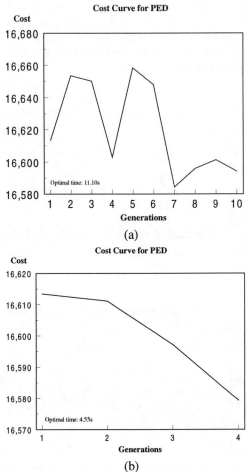

Figure 8 Evolutionary process of CGA and FCGA

10.6 Some Remarks

Firstly, this chapter presents the applications of a search methodology - Ant Colony Search algorithm - based on a distributed autocatalytic process. The individual ants are rather simple, however, the entirely colony foraging towards the bait site can exhibit complicated dynamics to result in very attractive search capability. The power of the algorithm comes from a process of dual inheritance. At the micro-evolutionary level there is a number of agents (ants), each can be described in term of a set of behavioural traits. At each step of the evolutionary process, each agent leaves a sign of its activity which changes the probability, with which the decisions will be made in future. At the marco-evolutionary level, individual agents are inherited from the 'queen' within the colony which generalises their experience. The results obtained clearly show the Ant Colony Search algorithm converges to the optimum solution through a autocatalytic process. The massive parallel agent co-operation makes the ants able to jump over the local optimum and to identify the right cluster easily, hence, a good solution can be found. Although the results in this paper are very encouraging, it should be observed that current research of Ant Colony Search algorithm in power system is still at a feasibility stage. More potentially beneficial work remains to be done, particularly in the areas of improvement of its computation efficiency. This is a common problem with biologically inspired algorithms such as genetic algorithms. Two approaches being currently investigated by the authors to achieve performance enhancement are: one is to develop advanced ant colony search algorithms incorporated with another intelligent technique and the second is to optimise parameters in the algorithm.

Secondly, the chapter presents AEC-GA. The biological mechanisms of bacteria provide additional insight into the effective implementation of advanced genetic operators. The proofreading system found in the high reproductive rate of bacteria have to be robust enough to remove replication errors. The artificial model is analogous to this. The ECO can be seen as one of the proofreading system. This uncovers the reason that AEC-GA hybrid is significantly better than the CGA as demonstrated in the ED applications. Clearly, the local improvement nature of the ECO certainly plays an important role in improving the dominant individual of the GA populations.

Thirdly, a FCGA has been described where two fuzzy controllers have been designed to adaptively adjust the crossover probability and mutation rate during the optimization process, based on some heuristics.

The three algorithms have been tested on a six-generator economic load dispatch problem. Compared with conventional GA and the Newton-Raphson method, the results reported have demonstrated the improved performances by the proposed algorithms. It is worth pointing out that more advanced hybrids will emerge to take advantages of each individual technique.

10.7 References

1. M Dorigo, V Maniezzo and A Colorni, The Ant System: Optimization by a Colony of Co-operating Agents, *IEEE Transactions on Systems, Man and Cybernetics*, 1995
2. V Chou, Y H Song, Ant colony-tabu approach for combined heat and power economic dispatch, *Proc 32nd UPFC,* 1997, pp.605-608

3. Y H Song, C S Chou, Advanced engineered-conditioning genetic approach to power economic dispatch, *Proc IEE - GTD*, Pt.C, Vol.144, No.3, 1997, pp.285-292
4. Y H Song, G S Wang, P Y Wang, A T Johns, Enviromental/economic dispatch using fuzzy controlled genetic algorithms, *Proc IEE - GTD*, Pt.C, Vol.144, No.4, 1997, pp.377-384
5. W D Potter, B E Tonn, R V Gandham, C N Lapena, Improving the reliability of heuristic multiple fault diagnosis via the engineered-conditioning-based genetic algorithm, *International Journal of Artificial Intelligence*, Vol.2, 1992, pp.5-23
6. J A Miller, W D Watterr, R V Grandham, C N Lapena, An evaluation of local improvement operators for genetic algorithms', *IEEE Trans. on Systems, Man, and Cybernetics*, Vol.23, No. 5, Sep/Oct 1993, pp.1340-1350
7. A J Wood, B F Wollenberg, Power generation, operation & control, *John Wiley & Sons*, New York, 1996

Chapter 11

INDUSTRIAL APPLICATIONS OF ARTIFICIAL INTELLIGENCE TECHNIQUES

A.O. Ekwue
National Grid Company plc, UK

11.1 Introduction

The electricity supply market in most countries of the world is being transformed from a regulated and monopolistic industry to a de-regulated utility with competition introduced. For example, in Europe, there has been early progress towards the restructuring and liberalisation of the electricity markets: England and Wales, Norway, Sweden, Portugal, Spain and Finland already have separate independent transmission companies. Hungary and Czech Republic are already considering restructuring their electricity supply industry. Considering other parts of the world, Argentina, Australia (Victoria), Canada (Alberta), Chile, India, New Zealand, Poland and Ukraine have all unbundled, stand-alone transmission systems [1].

Three main features of deregulation which will encourage the increasing use of artificial intelligence techniques for power systems include [2]:

- increase in the uncertainty of input data such as market forecasts;
- complexity of control centre operations with the increase in energy transactions;
- time pressures associated with competitive markets during operational planning and control phases.

Two techniques for heuristic searches are: genetic algorithms and simulated annealing. These methods solve conventional optimisation problems by "randomly generating new solutions and retaining better ones". Some disadvantages of these techniques include:

- It is difficult to define an appropriate stopping criteria so as to avoid unnecessary computational effort and at the same time not being too far away from the global optimum.
- An optimum solution is not always guaranteed.
- The parameters need to be tuned properly to obtain good performance.
- High dimensionality. For power system applications, the solution space is

usually large and could be computationally expensive.

Nevertheless, heuristic search methods have potential applications in power systems involving large, complex and non-linear optimisation problems such as unit commitment, economic dispatch, maintenance scheduling, generation expansion etc. Some of these algorithmic developments have been covered in previous chapters. As simulated annealing is still being developed for power system industrial applications, this chapter will concentrate mainly on the use of genetic algorithms, expert systems, artificial neural networks (ANN) and fuzzy logic.

The main emphasis of this chapter will be to present an overview of some of those applications that have already been implemented (or have the potential of being implemented) by industry as reported in the literature.

11.2 Algorithmic Methods Versus Artificial Intelligence Techniques

The main advantages of algorithmic methods include:

- optimality is mathematically rigorous in some algorithms;
- problems can be formulated to take advantage of the existing sparsity techniques applicable to large scale power systems;
- there exists a wide range of mature mathematical programming technologies, such as Mixed Integer Programming [3], Sparse Primal Linear Programming [4,8], Benders Decomposition [5-6], and Nonlinear Programming [7].
- fuzzy set theory has been applied to algorithmic optimisation to handle problem uncertainties [9,10] and
- a new optimisation strategy emulating the evolution process (Genetic Algorithm) has been used for improving the optimality search in an algorithm [11].

Despite the success of the algorithmic approaches described in the previous chapters, there remains a large class of problems that elude complete solution in a conventional setting. These problems:

- require the use of knowledge bases to store human knowledge;
- require operator judgement particularly under practical situations;
- require experience gained over a period of time;
- are characterized by network uncertainty, load variations etc

Hence the use of Artificial Intelligence techniques (Expert Systems, ANN, Genetic Algorithms, Fuzzy Logic etc) is being considered for power system applications [12].

11.3 Artificial Intelligence Techniques

11.3.1 HEURISTIC TECHNIQUE - GENETIC ALGORITHMS

Genetic Algorithms (GA) are general purpose optimisation algorithms as described by Holland [13]. They are distinguished from conventional optimisation by the use of concepts from population genetics to guide the optimisation search.
A GA technique operates on a population of individuals each representing a solution to the problem at hand. By a random process of cross-over and mutation, the GA combines features of the fittest individuals in a population to breed offspring; the aim is to preserve the desirable characteristics of individuals from one generation to the next generation offspring. The advantages of GAs over traditional techniques are:

- it needs only rough information of the objective function and places no restriction such as differentiability and convexity on the objective function;
- the method works with a set of solutions from one generation to the next, and not on a single solution, thus making it less likely to converge on local minima;
- the technique uses a population of solutions each represented by a bit string in which the flipping of a bit conceptually represents searching through a hyperplane in the solution space; the search effort is thus spread simultaneously in many regions of the search space.
- the solutions developed are randomly based on the probability rate of the genetic operators such as mutation and cross-over; the initial solutions thus would not dictate the search direction of a GA.
- GA can be implemented in parallel processing systems thus reducing computational time.

The use of GA has begun to be applied to the following power system optimisation problems:

- reactive power and voltage optimisation [11,14].
- Unit commitment [15,16]
- both unit commitment and economic dispatch problems [17]
- optimal selection of capacitors for radial distribution systems [18].

GA is equally suitable for short-term load forecasting when used in conjunction with neural networks and a combination of GA/fuzzy logic has been proposed for electric network planning [19]. Also, a hybrid GA/dynamic programming approach to optimal long-term generation expansion planning has been described in a recent paper [20]. In the latter example, the main attraction for employing this hybrid system is the fact that the GA finds the global optimal solution while the dynamic programming technique gets a local optimum. The suggested method has been applied to a practical long-term system with a 24-year planning period.

Unit Commitment is a highly constrained mixed-integer programming power system problem. According to Aldridge et al [21], two main approaches have been employed in applying GA to the unit commitment/economic dispatch problem:

- using a single GA for the entire scheduling period or
- decomposing the problem and using sequential GA at each time interval.

In their paper, four cases were examined:

- the basic GA;
- augmented GA with partitioning;
- augmented GA with heuristic dispatch and;
- both methods of augmentation

The results they presented showed that augmented GA with problem specific knowledge can generally enhance the performance of unit commitment/economic dispatch problems.

Transmission networks need to be maintained regularly so that utilities can optimise the use of their assets for greater efficiency. It therefore presents a large complex non-linear optimisation problem suitable for GA applications. In a recent paper, Langdon and Treleaven [22], described the scheduling of preventive maintenance of the south west peninsular region of the National Grid Company (NGC) electrical network. The combination of GA and hand coded heuristic and a genetic programming (GP) approach (using the same heuristics in the initial population) produced a low cost schedule for this network.

Chebbo and Irving [23] described a new transmission planning technique using the Deterministic Crowding GA. The optimal transmission network over an extended time horizon was identified based on the expected demand and generation pattern. The GA model was tested on a 23 bus network representing a simplified version of the NGC system to illustrate the principles of the method. The lines chosen were either single or double circuits at possible 275KV, 400KV and 750KV voltage levels. It was noted that due to the flexibility of GA, further modelling requirements can be included in the fitness function to enhance the design.

It must be stated that research into the applications of GA to industrial power systems is still in its infancy hence very few practical systems have been implemented. However, the preliminary results are encouraging. More research is also required into potential applications of GP. This is to exploit the fact that GP codes solutions as tree structured with variable length chromosomes whereas GA makes use of chromosomes of fixed length and structure [24].

11.3.2 EXPERT SYSTEMS

For an expert system, the main advantages are:

- it is permanent and is consistent;
- it can easily be transferred or reproduced and
- it can easily be documented.

Expert systems can be applied for monitoring and diagnosis, control, restoration, planning, security and load forecasting [2]. A typical real-time application for monitoring purposes was described by Ekwue et al [25] to emulate the functions of a junior control engineer. An example of fault diagnosis for transmission systems is described below (for more details see Esp and Ekwue, [26]).

Modern energy and management systems (EMS) must be capable of dealing with the numerous alarms received during fault and emergency conditions as a result of discrete and analogue data received by the SCADA (Supervisory Control and Data Acquisition) computers. These alarms can be generated for a variety of system conditions [27]:

- current limit exceeded;
- frequency deviation;
- voltage deviation;
- protection events;
- maloperation of protective equipment;
- non-functioning of remote controls etc.

In addition there may be switching indications (non-alarm) telemetry such as 'circuit breaker open' messages. To prevent the control engineer from losing important information within a mass of low-level alarms and events, there is the need to assist in their interpretation. One way of providing this assistance is to automatically diagnose the root cause of a received sequence of alarms etc.

Esp and Ekwue [26] presented the development and testing of a real-time Alarm Handling and Fault Analysis (AHFA) expert system facility within the NGC. AHFA was originally developed to assist Area Control Engineers by diagnosing faults on the power system from switching indications. These faults are typically generated during unusual weather system conditions, can generate many messages which indicate switching actions to be analysed by an engineer in the control centre.

The original objective of AHFA was to reduce the volume of data presented to the control engineer by diagnosing and summarising faults in place of the received telemetry. Recently, the control engineer requirements have been reconsidered and the preference is now to use the knowledge of diagnosed events (such as faults) to display the telemetry in groups associated with each event. In addition, the range of events which AHFA is required to diagnose has been expanded. The diagnostic function of AHFA, nevertheless, remained essentially unchanged.

Bann et al [28] described issues that must be addressed when integrating artificial intelligence techniques, particularly expert systems in the EMS. They reported that their developed artificial intelligence-EMS interface was simple, flexible and cost effective. Some of the practical implementations described include systems for a:

- a medium-sized utility from the Midwestern United States;
- large utility from the Midwestern United States;
- integrated restoration assistant in the OTS of a large Eastern United States utility.

Red Electrica in Spain has produced the following expert system facilities [29,30]:

- SEACON - an on-line expert system for contingency analysis and security enhancement;
- SEPDES - maintenance outage planning expert system;
- SEPTRE - voltage control and Mvar management system;
- SAR - restoration expert system

Liu and Tomsovic [31] described a prototype voltage/var control expert system (VCES), its main purpose being to prove the feasibility of the approach for this particular domain. VCES contained both knowledge of an empirical nature and knowledge supported by reasoning over cause-effect modelling. Subsequent to this research, an expert system was deployed at the control centre of the Portuguese national utility, EDP. The main problem was to solve the voltage/var problem within the appropriate context of the SCADA/EMS systems. There was also the need to deal with the problems associated with large, complex systems and computational efficiency in real-time environments.

The main advantages of the combined expert system and non-linear optimization functions at EDP were evaluated at multiple levels, both operational and organizational [32]. It was believed that the important results of this project were:

- real-time operation of a voltage/var expert system integrated with an optimization package;
- fewer, more localized and more pragmatic control actions;
- faster and reduced computational overhead; the expert function runs after each valid state estimation solution, typically on a 10-minute basis or event driven by topological changes;
- ability to infer the severity of the voltage control problem and adapt, as appropriate, the goals of both the knowledge based and the optimization function;
- demonstration of the feasibility and usefulness of efficient integration of different paradigms, i.e. algorithmic methods and AI related modules, in the present EMS environment of the control centres;
- systematization of the knowledge and data about the network resources, control deficiencies and operational practices, related to the specific domain of voltage / reactive management;
- redefinition of operational criteria concerning the coordination of controls, the management of reactive power reserves and the voltage profile monitoring;
- improvement on the accuracy of models and parameters, which lead to the dissemination of higher quality information to other departments in the utility, namely the planning department.

The experience gained with the EDP project gave an insight into the effective role of knowledge based systems in the EMS environment, and the evolutionary process of integration of artificial intelligence technology into the power system industry.

For NGC, a demonstrator was produced as a stand-alone version of the system at the EDP [33]. This comprised of a knowledge-based function, VCES (Voltage Control Expert System), which was developed to process a subset of alarms, those originated by voltage violations, under coordination with a multi-objective reactive power optimization function. This was built around the algorithmic kernel of the numerically robust MINOS (Modular Incore Nonlinear Optimization System) package. The rule base combined a generic engineering of the voltage/var problem with specific knowledge of the operational practices for the utility's network and its equipment models. The rules for the selection and implementation of available controls were based not only on the numerical values of sensitivities (topology condition), but also on such practical considerations as the usefulness of the controls, for example, through the evaluation of the available reactive reserve (load/generation condition), i.e. the control margin at each operating point.

Other practical expert system applications are described in references 34 and 35.

11.3.3 ARTIFICIAL NEURAL NETWORKS (ANN)

ANN technology and its characteristics have been reported extensively in the literature and the main advantages are: it

- is fast hence suitable for real-time applications;
- is robust and adapts to the data hence useful for both short-term and long-term load forecasting particularly if massive amount of historical data is available;
- possesses learning ability and it is appropriate for non-linear modelling. These advantages suggest their use for security monitoring (particularly for classification purposes) and control. Short et al [36] reported on an ANN-based voltage collapse investigation; the multi-layered feedforward perceptron was trained by a backward error propagation algorithm. The approach was based on the minimum singular value method. Though the neural network training is generally computationally expensive, the method described in this paper took negligible time to evaluate voltage stability once the network had been trained. Other applications are for real-time generator exciter control and intelligent relays; alarm processing and monitoring and fault diagnosis for networks, substations and equipment [2].

Despite these advantages of ANN method, some of the disadvantages of using this technology include:

- it may be difficult to scale ANN to solve realistic sized power system problems;
- selection of the optimum configuration - (and so the possibility of using GA to assist is being investigated);
- convergence difficulties;
- the choice of training methodology and the size of the training data - the use of GA in training are being explored;
- the 'black-box' representation of ANNs - they lack explanation capabilities and so decisions are not auditable;
- the fact that results are always generated even if the input data are unreasonable.

A general appreciation of the basic concepts of ANNs and an insight into how these networks can be employed to solve complex power system problems have been covered in recent tutorial papers by Aggarwal and Song [37,38]. Several

ANN practical implementations have also been described in the literature [19,39-41] but only few of such applications will be described here.

Highley and Hilnes in [42] described an ANN-based load forecasting case study for the Associated Electric Cooperative, Inc (AECI). AECI supplies power to rural electric cooperatives serving most of Missouri and parts of Southern Iowa. Conventional techniques based on regression and pattern matching algorithms did not perform very well as they did not possess the ability to adapt to rapidly changing weather conditions. In their study using ANN, they were able to reduce the RMS error forecasts from 5% to about 2%.

It has been stated in [19] that several utilities particularly in the USA have adopted the ANN-based load forecasting technique with improved prediction accuracy. Hannan et al [43] have applied multi-layer perceptrons to NGC's cardinal point forecasting. However, Majithia et al [44] believe that more research is required before a standard methodology for constructing a reliable ANN model is developed.

The philosophy behind contingency ranking, as is used in several EMS, is to predict the impact on the power system of various line/generator outages without actually performing full AC load flow analyses. However, existing contingency ranking methods suffer from either masking effects or slow execution despite the advances in computer technology. This is a heavy computational burden for contingency analysis. Hence the need to explore other alternative approaches such as using ANN. A number of methods for steady and dynamic security assessment have been proposed in [45-48].

One of the challenges for an on-line dynamic security assessment facility is the need for timely solutions. In an EPRI project, ABB Systems Control [48] investigated the integrated use of artificial intelligence techniques (rule-based expert systems, neural networks) and conventional analytical methods to achieve this objective. In an NGC sponsored project at the University of Bath, the results on the use of transient stability screening for the full NGC network was encouraging [49].

Finally, it has been suggested that the complementary methodologies of Expert Systems (ES) and Neural Networks might be combined to advantage in a hybrid system such as for voltage collapse monitoring [36]. In the hybrid scheme described, the ES carries out high-level monitoring, diagnosis, or planning, whereas the ANN is used to evaluate local problems which would otherwise involve great analytical complexity.

11.3.4 FUZZY TECHNIQUES

Fuzzy set theory allows a certain level of ambiguity in analysis [2] and can handle imprecise, vague or 'fuzzy' information. It has widely been applied for real-time control, operations, operational planning and planning particularly for non-linear problems. It is suitable where several conflicting objectives are involved. Fuzzy set theory has been applied to algorithmic optimisation to handle problem uncertainties [9,10].

Fuzzy logic has also been combined successfully with conventional optimisation or other artificial intelligence techniques. For example, in conjunction with expert systems and ANN. To illustrate this hybrid approach further, Bell et al [50] combined fuzzy set theory and expert systems for transmission line overload alleviation.

A general introduction to fuzzy logic has been described by Song and Johns [51]. An excellent literature survey of fuzzy set theory in power systems has been published by Momoh et al [52] with 107 references.

Finally, according to the UNIPEDE paper [19], fuzzy logic:

- is not suitable for alarm handling because this is a form of pattern recognition;
- has been applied at KEMA for security analysis in a project commissioned by the Dutch Electricity Generating Board. Also at the Israel Electric Corporation for contingency ranking based upon emergency control strategies;
- has been used for information retrieval at ENEL

11.4 Conclusions

The industrial applications of computational intelligence methods will likely increase following the on-going global deregulation of the electricity supply industry. The initiatives usually start with a prototype development before the technology is fully exploited. For example, Leahy et al [53] discussed an intelligent assistant for contract compliance in Coolkeeragh Power in Northern Ireland. They explained how knowledge-based systems can replace conventional methods for unit commitment to handle large contracts containing several thousand pages of information. It was reported that the system is now operational and with a government grant, it will be possible to integrate this facility in Coolkeeragh Power's day-to-day operations.

From the application of GA to large scale transmission network expansion planning on the North-Northeastern Brazilian network [54], it was found that interesting solutions were obtained with costs about 8.8% less that the best solution obtained by conventional optimisation techniques. Furthermore, this was 2.84% less than the best solution obtained by sequential simulated annealing and 1.13% less than that from a parallel simulated annealing. Also by adopting the constraint governed GA algorithm proposed by Hawken and Laughton [55], there could be improved clustering of results towards the optimum with reductions in computational requirements, particularly for optimum power flow related problems. These are encouraging results.

An exhaustive review of this subject cannot be claimed. However, the intention of this chapter is to illustrate the implementations and potentials of artificial intelligence applications to industrial power systems. However, in the light of the study carried out and presented here, the following future requirements may be concluded:

- despite remarkable advances in mathematical optimisation technology, conventional methods have yet to achieve fast and reliable real-time control applications; considerable efforts are required to avoid mathematical traps such as ill-conditioning and convergence difficulties;
- there are few industrial implementation of GA systems but the future is very promising. The parallelisation of GA for large power systems need to be explored.
- the application of GP should equally be encouraged. This is to exploit the fact that GP codes solutions as tree structure with variable length chromosomes whereas GA makes use of chromosomes of fixed length and structure [24].
- the experience of prototype expert systems applied to voltage control demonstrates that:
- knowledge-based systems can enhance the capabilities of a power system in handling reactive power control;
- knowledge-based development methodologies are relevant to reactive power control.
- for ANN, the selection of an appropriate training set to solve practical power networks remains generally a challenging research area [19];
- the application of hybrid systems is a novel development which represents a future trend in research.
- the application of fuzzy set theory to power systems is a new field for research but more attention should be concentrated on the suitable problems [52].

11.5 Acknowledgement

The work described here has been taken from several sources for which the author is very grateful.

11.6 References

1. Jefferies, D G (1997), *Developing the Highways of Power*, IEE Power Engineering Journal, 5-9
2. Dabbaghchi I, Christie R D , Rosenwald G W and Liu C C (1997), *AI Applications in Power Systems*, Expert Intelligent Systems & their Applications, Vol **12(1)**, 58-66
3. Aoki K, Fan W and Nishikori A, (1988), *Optimal VAR Planning by Approximation Method for Recursive Mixed-Integer Programming*, IEEE Trans on Power Systems, Vol **3(4)**, 1741-1747
4. Iba K, Suzuki H, Suzuki K I and Suzuki K, (1988), *Practical Reactive Power Allocation/Operational Planning Using Successive Linear Programming*, ibid, 558-566.
5. Deeb N and Shahidehpour S M, (1990), *Linear reactive Power Optimisation in a Large Power Network Using the Decomposition Approach*, ibid, Vol **5(2)**, 428-438.
6. Granville S, Pereira M and Monticelli A, (1988), *An Integrated Methodology for VAR Sources Planning*, ibid, Vol **3(2),** 549-557.
7. Sun D I, Ashley B, Brewer B, Hughes A and Tinney W F, (1984), *Optimal Power Flow by Newton Approach*, IEEE Trans on Power Apparatus and Systems, Vol **103**, 2864-2880
8. Alsac O, Bright J, Prais M and Stott B, (1990), *Further Developments in LP-Based Optimal Power Flow*, IEEE Trans Power Systems, Vol **5(3)**, 697-711.
9. Abdul-Rahman K H and Shahidehpour S M, (1993), *A Fuzzy Based Optimal Reactive Power Control*, (1993), ibid, Vol **8(2)**, 662-670.
10. Tomsovic K, (1992), *A Fuzzy Linear Programming Approach to the Reactive Power/Voltage Control Problem*, ibid, Vol **7(1),** 287-293.
11. Iba K, (1994), *Reactive Power Optimisation by Genetic Algorithm*, ibid, Vol **9(2)**, 685-692.
12. Warwick K, Ekwue A O and Aggarwal R K (editors) - 1997, *Artificial Intelligence Techniques in Power Systems*, IEE Power Engineering Series (22), London
13. Holland J H, (1975), *Adaptation in Natural and Artificial Systems*, University of Michigan Press
14. Haida T and Akimoto Y, (1991), *Voltage Optimisation using Genetic Algorithms*, Third Symposium on Expert System Applications to Power Systems, Tokyo, Japan, 375-380
15. Dasgupter D and McGregor D R (1994), *Thermal unit commitment using genetic algorithm*, IEE Proceedings Part C Vol **141 (5)**, 459-465
16. Orero S O and Irving M R, (1997), *Large scale unit commitment using a hybrid genetic algorithm*, Electrical Power and Energy Systems, Vol **19 (1)**, 45-55

17 Hassoun, M H and Watta P (1994), *Optimisation of the unit commitment problem by a coupled gradient problem network and by a genetic algorithm*, Project 8010-29, Electric Power Research Institute
18. Sundhararajan S and Pahwa A, (1994, *Optimal selection of capacitors for radial distribution systems using genetic algorithms*, IEEE Trans on Power Systems, Vol. **9** (**3**), 1499-1507
19. UNIPEDE CORECH Working Group (1997), *Neural networks, fuzzy logic, and genetic algorithms in the electricity supply industry: models and applications*, Engineering Intelligent Systems, Vol **5** (**3**), 127-155
20 Park YM, Park J B and Won J R, (1998), *A hybrid genetic algorithm/dynamic programming approach to optimal long-term generation planning*, Electrical Power and Energy Systems, Vol. **20** (**4**), 295-303
21 Aldridge C J, McDonald J R and McKee S, (1997), *Unit commitment for power systems using a heuristically augmented genetic algorithm*, Proc of the IEE Genetic Algorithms in Engineering Systems: Innovations and Applications, University of Strathclyde, 433-438
22. Langdon W B and Treleaven P C (1997), *Scheduling maintenance of electrical power transmission networks using genetic programming*, in Warwick K, Ekwue A O and Aggarwal R K (editors), Artificial Intelligence Techniques in Power Systems, IEE Power Engineering Series (**22**), London 220-237
23 Chebbo H M and Irving M R, (1997), *Application of Genetic Algorithm to Transmission* Planning, Proc of the IEE Genetic Algorithms in Engineering Systems: Innovations and Applications, University of Strathclyde, 388-393
24 Willis M J, Hiden H G, Marenbach P, McKay B and Montague G A (1997), *Genetic programming: an introduction and survey of applications*, ibid, 314-319
25 Ekwue A O, Esp D G, Macqueen J F, Vaughan B W, Short M J, Knight U G and Liernan L (1995), *System Monitoring Expert Systems*, International Journal of Engineering Intelligent systems for Electrical Engineering and Communications, Vol **3**(**2**), 87-94
26 Esp D G and Ekwue A O, (1996), *Real-time fault diagnosis for transmission systems*, Proc of UKACC International conference on CONTROL'96, University of Exeter, 1413-1417
27 Hasan K, Ramsay B and Moyes I (1994), *Object oriented expert systems for real-time alarm processing*, Electric Power Systems Research Journal (**3**), 69-82
28 Bann J, Irisarri G, Kirschen D , Miller B and Mokhtari S, (1996), *Integration of artificial intelligence applications in the EMS: issues and solutions*, IEEE Trans on Power Systems, Vol **11**(**1**), 475-482
29 Criado R, Matauco D, Lasheras F, Fernandez J L, Basagoti P and Serna J, (1992), *SEACON: an on-line expert system for contingency analysis and corrective solutions in transmission*, International Journal of Electrical Power and Energy Systems, Vol **14**(**4**), 303-309

30. *SAR: Sistema de Apoyo a la Reposicion Restoration Expert System*, CIGRE SC-39 Power System operation and Control, colloquium paper 4.4, 267-277
31. Liu CC and Tomsovic K, (1986), *An Expert System assisting decision-making of reactive power/voltage control*, IEEE Trans on Power Systems, Vol **1** (**3**), 195-201,
32. Barruncho L M, Sucena Paiva J P, Liu C, Pestana R and Vidigal A, (1992), *Reactive Management and Voltage Monitoring and Control*, Int. J of Electrical Power and Energy Systems. Vol **14** (**2-3**), 147-157,
33. Barruncho L M, Liu C, Ekwue A O, Esp D G, Macqueen J F and Vaughan B W, *(1994), Application of a knowledge based system for voltage monitoring and control on the NGC system,* Proc of the IEE CONTROL'94 Conference, 146-152
34. Special issue on *Expert System Applications in Power Systems*, (1992), Int. Journal of Electrical Power and Energy Systems, Vol **14** (**2-3**), 69-255
35. Special issue on *Knowledge-Based Electrical Power Systems*, (1992), Proc of the IEEE, Vol **80** (**5**), 659-778
36. Short M J, Hui K C, Macqueen J F and Ekwue A O, (1994), *Application of Artificial Neural Networks for NGC Voltage Collapse Monitoring*, paper no: 38 - 205 CIGRE Symposium, Paris.
37. Aggarwal R K and Song Y H, (1997),*Artificial neural networks in power systems: part 1 general introduction to neural computing*, IEE Power Engineering Journal, Vol **11** (**3**),129-134
38. Aggarwal R K and Song Y H, (1998), *Artificial neural networks in power systems: part 2 types of artificial neural networks*, IEE Power Engineering Journal, Vol **12** (**1**), 41-47
39. Proceedings of the International Conference on Intelligent System Applications to Power Systems (ISAP'94), 1994
40. Niebur D et al, *Neural Network Applications in Power Systems*, (1993), International Journal of Engineering Intelligent Systems, Vol. 1(3), 133-158
41. Niebur D and Dillon T S (editors), (1996),*Neural Network applications in power systems*, CRL Publishing Ltd, UK
42. Highly D D and Hilmes T J, (1993), *Load Forecasting by ANN*, IEEE Computer Applications in Power, Vol **6** (**3**), 10-15
43. Hannan J M, Majithia S, Rogers C and Mitchell R J (1996), *Implementation of neural networks for forecasting cardinal points on the electrical demand curve*, 4th IEE International Conference on Power System Control and Management, London, 160-166
44. Majithia S, Kiernan L and Hannan J M, (1997), *Intelligent systems for demand forecasting*, **in** Warwick K, Ekwue A O and Aggarwal R K (editors), Artificial Intelligence Techniques in Power Systems, IEE Power Engineering Series (**22**), London, 259-279
45. Lo K L, Peng L J, Macqueen J F, Ekwue A O and Cheng D T Y, (1995), *Application of Kohonen Self-Organising Neural Network to Static Security Assessment*, 4th IEE International Conference on ANN, Cambridge, 387-392

46. Lo K L, Peng L J, Macqueen J F, Ekwue A O and Cheng D T Y, (1998), *Fast Active Power Contingency Ranking Using a Counterpropagation network*, to appear in the IEEE Trans on Power Systems
47. Sobajic D J and Pao Y H, (1989), *ANN based dynamic security assessment for electric power systems*, IEEE Trans on Power Systems, Vol. 4(1), 220-226
48. Kumar A B R, Brandwajn V and Ipakchi A, (1991), *Applying artificial intelligence techniques to DSA*, EPRI publication AUTOMATION, **5**
49. Edwards A R, (1995), *Detection of Dynamic Instability in Power Systems using Connectionalism*, Ph.D thesis, University of Bath
50. Bell K R W, Daniels A R and Dunn R W, (1996), *A Fuzzy Expert System for Overload Alleviation*,Proceedings of the 12th PSCC, Dresden, 1177-1183
51. Song Y H and Johns A T, (1997), *Applications of fuzzy logic in power systems: part 1 general introduction to fuzzy logic*, IEE Power Engineering Journal, Vol **11** (**5**),219-222
52. Momoh J A, Ma X W and Tomsovic K , (1995), *Overview and Literature Survey of Fuzzy Set Theory in Power Systems*, IEEE Trans on Power Systems, Vol **10** (**3**), 1676-1689
53. Leahy D G, Wallace J G, Mulvenna M D and Hughes J G, (1997), *Knowledge-based assistance for contract compliance*, IEEE Expert Intelligent Systems & their Applications, Vol **12**(**2**), 58-64
54. Gallego R A, Monticelli A and Romero R, (1998), *Transmission System Expansion Planning by an extended genetic algorithms*, IEE Proc - Generation, Transmission and Distribution, Vol **145** (**3**), 329-335
55. Hawken A D and Laughton M A, (1996), *Constraint Governed Genetic Algorithm*, Proceedings of the 12th PSCC, Dresden, 506-512.